Trading in oil futures

SECOND EDITION

SALLY CLUBLEY

Nichols Publishing Company
NEW YORK

Published in the United States of America by
Nichols Publishing Company
PO Box 96, New York, NY 10024, USA

Published in Great Britain by Woodhead-Faulkner Limited,
Simon & Schuster International Group,
Fitzwilliam House, 32 Trumpington Street,
Cambridge CB2 1QY, England

First published 1986
Second edition 1990

Library of Congress Cataloging-in-Publication Data

Clubley, Sally.
Trading in oil futures / Sally Clubley. -- 2nd ed.
p. cm.
ISBN 0-89397-360-2: $29.95
1. Petroleum industry and trade. 2. Futures market.
3. Speculation. I. Title
HG6047.P47C57 1990
332.64′42282 -- dc20 89-70953
CIP

Designed by Geoff Green
Typeset by Goodfellow & Egan (Phototypesetting) Ltd, Cambridge
Printed in Great Britain by BPCC Wheatons Ltd, Exeter

Contents

Preface

This book is intended as an introduction to the energy futures markets around the world. It is primarily aimed at those new to the futures markets, but will also serve to show those in other sectors of the futures markets some of the ways in which the oil industry is using the markets. It looks briefly at the histories of both the oil and futures industries and how changes in the former made a joint development inevitable. The rapid development of the oil spot markets during the 1970s and the erosion of the power of the major oil companies and, later, the oil-producing countries combined with the high volatility of oil prices led to a need for futures markets.

The book looks at all the futures markets currently trading around the world, and some of those proposed for the near future, and the ways in which they can be used by those involved in the oil industry, whether producers, traders or consumers, to offset risk and improve cash flow. Without an understanding of the basic principles of the futures markets it is difficult to achieve any degree of success in using the markets. This understanding normally develops over time and includes an introductory trading period with very strict controls.

Tight controls will, of course, be maintained throughout any trading activity. It is relaxation or absence of these controls which has led to most of the major losses immortalised in popular futures market history. Although it may be tempting to blame the markets themselves, it must be remembered that the energy futures markets have been established to serve the oil industry not to threaten it.

Judicious futures trading is, as many traders are now finding, a useful adjunct to physical trading and can enable almost any company to create more imaginative and useful trading strategies.

Sally Clubley
September 1989

1

The development of the world oil industry

Origins of the modern oil industry

Until the middle of the last century the world had little interest in crude oil. Then the first commercial oil discovery was made, and the beginnings of the refining industry were founded. Of course, the existence of crude oil had been known for centuries; it had been found seeping out of rocks in biblical times and afterwards, though its usefulness, particularly as a fuel, was never really appreciated. It was treated in the same way as tar and primarily used for water-proofing purposes.

The real origins of the modern oil industry are in Pennsylvania, where Edwin Drake made the first commercial discovery of crude oil in 1859. He is generally acknowledged to be the first person specifically to drill for, and find, crude oil. Very quickly he was followed by others, both in Pennsylvania and elsewhere. This first oil rush led swiftly to the first oil glut and a fall in price more dramatic than any seen since, with prices dropping from $20 per barrel in 1860 to 20 cents per barrel only a year later.

Oil refining was, in fact, a technique already in use when Pennsylvania's first oil was drilled. Some ten years earlier, a plant had been built in Scotland to process the shale oil seeping into coal mines in the area. But the small-scale project failed to attract any real interest and refining can really be said to have started with the treatment of the American oil. Initially, of course, uses had to be found for the new oil products. Oil lamps were soon prevalent in oil-producing regions and lubrication developed as another major application in the early years. Within a very short time, oil had begun to establish its almost unassailable position as an essential fuel source and raw material in the industrialising world.

But the political and economic upheavals that oil would cause a hundred or so years later were still unthought of as the race to find more black gold continued.

By the turn of the century the United States and Russia were already established as the world's leading oil producers, a position they still hold today, thus beginning a domination that was to last until the large producers of Venezuela and the Middle East combined forces in the mid-twentieth century. In 1908 oil was first discovered in Iran but it was not until many years later that production in the Persian Gulf (then, as now, the cheapest area of production in the world) got underway.

The major oil companies

Just before the turn of the century a race developed to capture the largest number of drilling rights in the Persian Gulf – an area rich in oil deposits, but lacking the technological ability to develop them. The runners in this race were the major oil companies, known as the 'Seven Sisters', of whom six still exist. These companies, despite intense competition and bitter arguments, operated effectively as a cartel dominating the oil industry until the 1960s, and still have a tremendous influence.

Five of the Sisters were American, three of them (Exxon, Mobil and Chevron) the offspring of one corporation – John D. Rockefeller's Standard Oil Company. Rockefeller was one of the first people to recognise the importance of integrating oil company activities and keeping control of the oil at the drilling stage, through refining and then delivery into the consumers' oil tanks. The company dominated the US oil industry and, by implication, the world oil markets, but it was eventually disbanded by legislation in 1920. The group splintered into around forty separate companies, each operating in one state. Of these, those in New Jersey (Exxon), New York (Mobil) and California (Chevron) became the most important. The other two US Sisters, Gulf, since taken over by Chevron, and Texaco, both began life in Texas, which still plays a vital part in the US oil industry with a large proportion of the refining capacity, many pipeline terminals, landing ports and trading companies. Rockefeller had been excluded from Texas from the beginning and never managed to gain a foothold in the state where North America's greatest oil reserves to date were found in the late nineteenth century.

The remaining two Sisters were both European – British Petroleum and Royal Dutch/Shell. The latter was formed by the amalgamation of two companies: one Dutch, one British. The growth of both these Sisters was based on oil reserves far from home – BP in Iran and Russia, and Shell in Venezuela.

These seven companies completely monopolised the industry until the 1960s; drilling for, producing and refining the crude, distributing the products and, finally, retailing them to the consumer. As the oil potential of the Middle East became apparent, the companies formed a series of consortia (after much battling between themselves) to negotiate with the local governments and rulers and arrange production deals. The Middle Eastern governments at the time had no complaints; they were simply happy to see their income growing.

Outside the seven, the only company that achieved any degree of success was the French national oil company, Companie Français de Pètrole (CFP). This was admitted into consortia which were involved in Iran and Iraq, though not at the very beginning. It was not until CFP discovered oil independently in Algeria that it was really ranked as a major oil company.

From the beginning the industry was faced by a major problem: transportation. (Shell, in fact, developed from a shipping company but found itself struggling to survive until it merged with Royal Dutch.) Outside North America oil was being produced many miles from the areas of demand – in Venezuela, Mexico, the Persian Gulf and Russia. There existed a number of agreements between the Seven Sisters to exchange oil, in order to prevent the transportation costs becoming too much of a burden. All posted (i.e. official) crude oil prices in the early twentieth century were based on a theoretical price in the Gulf of Mexico *plus* transport costs, so these exchanges could be quite advantageous to the companies concerned.

The detailed history of the world's oil industry has been well chronicled elsewhere but some brief notes are necessary to an understanding of the current state of affairs and likely future developments.

The oil industry in the twentieth century

In the early part of this century it seemed that all sides of the industry were reasonably content with the way things were going. The producer countries were well paid for the use of their resources; the oil companies were all enjoying high profits and had plenty of oil reserves. After the Second World War the battle for a share of the market became the most pressing problem for the oil companies – oil demand was booming and no one wished to get left behind in the rush for expansion.

But already the signs of change were, in retrospect, becoming apparent. In the Middle East, the Gulf states and Iran had been using the income from oil to send their young men to Europe and the United States to receive a better education in technological subjects, politics and economics. When they returned they began to question the situation whereby the producing governments had virtually no control over their own resources. The governments had, however, initiated costly and extensive development programmes which were dependent on continuing wealth and were wary of damaging their relationships with the oil companies.

The first move was made by Venezuela, which passed a law in 1948 requiring the oil companies to hand over 50 per cent of their profits. The companies, realising they had little alternative, agreed and the idea quickly spread to the Middle East, where it was taken up, with similar results, first by Saudi Arabia and later by others.

This resentment of the oil giants spread, especially with the gradual emergence of independent oil companies set up either by individuals like J. Paul Getty or by consumer government agencies such as Agip in Italy. At around the same time, too, CFP made its Algerian discoveries. These independents began to make production agreements with the producer governments, offering better royalties than the Sisters. This suggested to the producers that they were not being paid as much as they could be, and they began to consider ways of swinging the advantage towards themselves. The first concrete move came in the form of an alliance, unlikely as it may seem today, between Saudi Arabia and Iran. The terms were very loose, amounting to little more than a vague co-operation agreement, and had virtually no effect on anything. Meanwhile, Venezuela was continuing an effort started some years earlier to persuade the producers to get together to form some sort of combined opposition to the multinational companies.

Throughout the 1950s, however, the status quo was maintained, with most of the producers coming to better agreements with the companies but otherwise leaving operations alone. One notable exception to this was Iran, which nationalised its oilfields in 1951. This move was followed by a Western boycott of Iranian oil; a small sacrifice for the companies (apart from BP, which was very reliant on Iranian oil) because there was once more a glut on the world oil markets. Two years later, Western governments intervened to bring down the revolutionary regime in Iran and re-establish a climate in which the Iranian oil industry could restart operations. Although the glut made the companies somewhat reluctant to increase production, the Iranians were in severe financial difficulties and it was essential, politically, that contact be made. At this stage the BP monopoly in Iran was broken and the Seven Sisters formed a consortium with CFP to continue the development of Iranian oil. Despite the fact that one company had suffered, it now seemed that control of the industry was firmly back in the hands of the seven major oil companies.

The next event of importance was the Suez crisis of 1956, as a result of which the Arab states imposed an oil embargo on the West. Although the total world oil supply was hardly affected, because production was increased elsewhere, irreparable damage was done to the relationships between the producers and the oil companies, particularly the two European ones. Perhaps the main feature of the embargo was that it showed that the producers could act together when sufficiently aroused. In the aftermath of Suez, several of the smaller independents reached still 'better' terms with the producers, and further damage was done to the Seven Sisters.

Throughout this time, the oil glut was continuing. While it was in nobody's interests (except those of the consumer, who had little say in things) to reduce the price of oil, it soon became inevitable that this would happen. In 1959 the crude oil price was cut for the first time this century. The inevitability of the move meant that there was little real protest from the producers; but a meeting of the Arab Petroleum Congress quickly followed. The meeting produced no firm action. But, when the companies imposed a second price cut the following year, the Congress met again, this time with Venezuela in attendance, and the Organisation of Petroleum Exporting Countries (OPEC) was formed by Venezuela, Saudi Arabia, Iran, Iraq and Kuwait.

However, there was no dramatic action from OPEC and everything carried on much as it had before. The group pledged co-operation to avoid 'unnecessary fluctuations' in the price of oil; but even so, the coming changes in the structure of the industry were still virtually imperceptible.

Oil demand was booming, production was increasing and incomes following suit: there was no reason for any conflict. Prices did not increase sharply, gradually rising from $1.20 per barrel in 1960 to $1.80 a decade later. But the higher production levels kept the producers' incomes rising steadily. The only worry facing the industry was whether the oil reserves would last until the end of the century, but even this was of no real concern – reserves were increasing as new areas were explored and drilled, and technology was improving, allowing more oil to be extracted from every well and enabling previously impossible reserves to be developed.

Perhaps the most significant change during the 1960s was the increasing part played by the major US oil companies in the European market, following the imposition of import controls by the US government in 1957. The US Sisters and independents had to market all their Middle Eastern, North African and other foreign oil outside the USA. The only real market available was Europe, and oil companies launched into fierce advertising battles to increase market share and develop brand loyalty in the consumer, particularly in the fast-growing gasoline market. The degree of success achieved in this latter objective is perhaps surprising, but many motorists were impressed by the promotional offers and the advertising campaigns – the most successful of which was Exxon's 'Put a tiger in your tank'. By the time the US import controls were lifted, the major oil companies had established lucrative European markets, where they remained active until the early 1980s, when one or two, notably Gulf, withdrew altogether from Europe.

Since the Suez crisis, when the oil boycott had been successfully executed, albeit with little effect, there had been some concern about an orchestrated move by the oil producers to block exports to the West for political or economic reasons. But, by 1970, when the next move was made, many had come to believe that any concerted action was unlikely – the producers were thought to be enjoying their increasing wealth too much.

The oil price rises of the 1970s

In 1970, however, the Libyan government imposed reduced production levels on Occidental Petroleum, an independent oil company totally reliant on Libyan oil. The company's production was cut back by almost half, forcing it to agree to higher posted prices and an increased royalty for the Libyan government. At the semi-annual OPEC meeting at the end of that year, the ministers called for a 55 per cent royalty agreement for all member countries. Negotiations with the oil companies resulted in an acceptance of OPEC's terms, on the condition that, apart from an agreed increase to allow for inflation, there would be no new demands for five years. In 1973, the OPEC countries decided to impose a 70 per cent increase in prices. The announcement came during a boycott on oil supplies to the United States and The Netherlands following the Yom Kippur war. World supply was short and prices rose – by the end of 1973 the posted price for Arab light crude was $11.65 and the spot price more than $20 per barrel. Prices on the spot market, which at that stage handled only small amounts of oil left over from term contracts, had never before risen above the official price. Although the spot market, then as now, tended to give a somewhat exaggerated picture, it indicated the industry's fears.

During the same year Saudi Arabia, the largest producer within the cartel, obtained an improved participation agreement with the oil companies, not only for itself, but also for the smaller Gulf producers – Kuwait, Abu Dhabi and Qatar. The agreement gave the producers a 25 per cent equity in production, rising to more than 50 per cent in 1982. Libya had achieved a similar result by nationalising its oilfields and Iraq had taken over 100 per cent equity. In fact, further pressure from the producers led to a much faster takeover than originally planned, with several achieving 100 per cent equity by the end of the 1970s.

Consumers responded to the sharp price rises of 1973 by cutting back, and demand slumped. Most continued to blame the oil companies for the increases, not as yet realising that they had all but lost control of pricing. These cutbacks were, however, short-lived. When the oil embargo was lifted in the middle of 1974, demand quickly began to rise again. This trend continued, even through the steady price rises of the next few years. The failure of the West to anticipate the Iranian revolution and the fall of the Shah at the beginning of 1979 has been well documented. Quite why the warnings of so many informed sources were ignored for so long remains something of a mystery. But when it came, the revolution, and the subsequent war between Iraq and Iran a year later, was to bring about the second oil crisis of the 1970s. Immediately after the revolution oil production in Iran dropped sharply. Prices rose on the spot market, but it was some months before any increase in posted prices was imposed. But the real harm was not done until the beginning of the war. Iranian production had fallen from a peak of over

6 million barrels per day (m.b.d.) in 1974 to 5.7 million barrels per day in 1978 as the internal unrest grew. By the end of 1979 the country was producing little more than 1 m.b.d. and Iraqi output fell from over 3.5 m.b.d. to less than 1 m.b.d.

Although significant areas of production were being developed elsewhere, the industry was still walking a tightrope between supply and demand, and was unable to cope with such a large cutback in supply. Product and crude prices on the European and US spot markets more than doubled during the course of 1979. Day after day, the price just continued to rise, and any trader managing to get hold of oil was sure to make a profit. All logistical factors were forgotten as the scramble for oil continued.

In time, however, the sharp escalation in prices had a corresponding, though opposite, effect on demand. 'Free world' oil demand peaked at 51.2 m.b.d. in 1979, falling to 48.6 m.b.d. the following year, and continued to drop, reaching 44.8 m.b.d. in 1983, the lowest level for ten years. Cause and effect may be difficult to determine, but the world recession following the high-inflation, high oil-cost years of the 1970s is undoubtedly the main reason for the fall. But the spin-offs – increased conservation and the switch to alternative forms of energy – have played a strong part and are to some extent irreversible. The causes remained difficult to measure as changes in industrial technology and improved energy conservation meant comparisons were impossible even when the world moved out of recession in the mid-1980s. One survey suggested that heating-oil demand per house for space heating fell by around 40 per cent in Germany from its 1979 high, largely because of more efficient heating systems and better insulation.

After the 1979 crisis the OPEC members decided to introduce price differentials for their crude oils. Changing demand patterns had made the lighter North African crudes, for example, very much more attractive than the heavier Gulf crudes, which produced less gasoline and more fuel oil. Arab light crude, with the largest volume, was chosen as the marker, with premiums of up to $9 per barrel being paid for some of the better crudes, and discounts being applied to the very low-quality ones.

It was these differentials, rather than the actual price of oil, which brought OPEC close to collapse in the early part of 1983. A year earlier, the benchmark price had been lifted to $34 per barrel and the maximum differential cut to $3 in an attempt to halt the slide of oil prices seen in the spring of 1982, as production continued at high levels and demand kept falling. For a while it looked as though this attempt might have been successful.

Meanwhile, crude-oil producers outside OPEC were gaining in importance, most notably the United Kingdom, Norway and Mexico. In 1973 three of the world's top five oil producers were in OPEC, but ten years later the United Kingdom and Mexico had displaced Iran and Libya, leaving only Saudi Arabia in the top five. In the first quarter of 1982 non-OPEC production exceeded OPEC output for the first time in 20 years. OPEC was becoming the swing producer. Table 1.1 gives a breakdown of world oil production figures between 1970 and 1988.

All these factors contributed to the worsening problems facing OPEC when it met in December 1982 and, for the first time in its history, failed to reach agreement on pricing. The meeting was adjourned until January, when the same thing happened again. The price of oil was becoming of secondary importance by this stage; differentials and production quotas were the problem. In early 1982 OPEC had sought to stabilise prices by imposing a maximum combined output level of 17.5 m.b.d. Although it wished to retain this overall ceiling, there was pressure for some reallocation of individual quotas.

Finally, in March 1983, the producer countries came to an agreement, cutting the benchmark to $29.00 per barrel and restricting differentials to $1.50. Production quotas were also agreed and, to the surprise of many in the industry, prices began to stabilise. Spot prices for Arab light, the marker crude, had fallen from $33.50 per barrel in September to $28.20 in March; but, by September 1983, they had moved back up to official levels.

The stability was helped by positive signs of some upturn in demand in the United States, which began to move out of the recession more quickly than Europe or Japan. Fourth-quarter demand was forecast to increase throughout the industrialised world and OPEC began to step up production in anticipation. By October 1983 there were few signs that this was happening and there was some unease creeping back into the industry.

Prices fell towards the end of 1983 and it appeared that another crisis was unavoidable when the oil industry packed up for Christmas. The weather, however, came to OPEC's rescue. North America was plunged into one of the coldest winters on record, with oranges freezing on the trees in Florida and refineries forced to close on the Gulf of Mexico. Oil demand leapt and large volumes of heating oil flowed across the

Table 1.1 World oil production (million barrels per day), 1975–88

	1975	1976	1977	1978	1979	1980
Saudi Arabia	6,970	8,525	9,235	8,315	9,553	9,990
Iran	5,385	5,920	5,705	5,275	3,175	1,480
Iraq	2,260	2,415	2,350	2,560	3,475	2,645
Kuwait	1,885	1,965	1,835	1,945	2,270	1,430
Neutral Zone	500	465	390	470	570	545
United Arab Emirates	1,695	1,945	2,005	1,835	1,830	1,700
Qatar	435	495	445	485	510	460
Venezuela	2,425	2,375	2,315	2,235	2,425	2,235
Ecuador	160	185	190	205	215	205
Indonesia	1,305	1,505	1,690	1,635	1,590	1,575
Libya	1,480	1,930	2,065	1,985	2,090	1,830
Algeria	1,020	1,075	1,150	1,230	1,255	1,120
Nigeria	1,785	2,065	2,085	1,895	2,300	2,055
Gabon	225	225	220	210	205	175
Total OPEC	27,530	31,090	31,680	30,280	31,465	27,445
United States	10,010	9,735	9,865	10,275	10,135	10,170
Canada	1,735	1,605	1,610	1,575	1,770	1,725
United Kingdom	30	240	765	1,095	1,600	1,650
Norway	190	280	275	350	385	525
Mexico	790	875	1,085	1,330	1,630	2,155
Total free world	43,855	47,470	49,125	48,855	51,280	47,945
China	1,490	1,675	1,880	2,090	2,130	2,125
USSR	9,935	10,525	11,055	11,595	11,870	12,215
Total centrally planned economies (CPEs)	11,830	12,605	13,425	14,200	14,495	14,810
Total world	55,685	60,075	62,550	63,055	65,775	62,755

Source: BP

Atlantic from Europe, supporting the market there. Stocks had been allowed to fall through the preceding autumn, because of the falling prices and unchanging demand, so the sudden surge in consumption had an exaggerated effect which lasted through the early part of the spring as stocks were rebuilt.

Later in the year the oil price began to fall again, but this time the OPEC producers were helped by the war between two of its members, Iran and Iraq. A large oil tanker, the *Yanbu Pride*, was attacked by Iranian ships off the coast of Saudi Arabia and the oil industry's fears of a cut-off in the oil supply from the Gulf were revived and prices leapt again. After a few weeks and some lesser attacks on shipping it became

1981	1982	1983	1984	1985	1986	1987	1988
9,985	6,695	5,225	4,760	3,565	5,150	4,360	5,255
1,325	2,410	2,465	2,195	2,215	1,905	2,310	2,275
895	1,010	1,105	1,225	1,440	1,745	2,090	2,600
965	705	900	985	920	1,250	1,075	1,340
375	315	310	405	340	345	390	320
1,495	1,245	1,175	1,200	1,280	1,480	1,650	1,730
425	340	310	425	340	355	340	360
2,180	1,965	1,875	1,875	1,730	1,845	1,775	1,925
210	200	240	255	275	280	165	150
1,680	1,415	1,345	1,410	1,335	1,400	1,320	1,315
1,220	1,135	1,110	1,105	1,105	1,045	1,000	1,055
1,035	1,045	980	1,075	970	1,060	1,040	1,070
1,440	1,285	1,235	1,385	1,475	1,465	1,290	1,365
150	155	155	150	155	160	155	170
23,380	19,930	18,425	18,470	17,215	19,555	19,030	21,145
10,180	10,200	10,245	10,505	10,545	10,230	9,945	9,750
1,545	1,485	1,515	1,645	1,815	1,840	1,695	1,775
1,835	2,125	2,360	2,580	2,655	2,665	2,555	2,360
505	530	660	755	835	910	1,000	1,150
2,585	3,005	2,950	3,015	3,015	2,750	2,875	2,855
44,525	42,075	41,545	42,865	42,645	44,645	44,025	46,320
2,035	2,050	2,135	2,300	2,515	2,630	2,675	2,735
12,370	12,430	12,520	12,450	12,150	12,560	12,745	12,705
14,850	14,985	15,145	15,235	15,130	15,660	15,870	15,875
59,375	57,060	56,690	58,100	57,595	60,305	59,895	62,195

apparent that there was no interruption to supply and the prices began to fall again.

Yet again, the autumn saw falling prices, with the British National Oil Corporation and Norway's Statoil forced to cut their contract prices in October. This move was followed by cuts in US domestic crude oils and a number of other smaller producers. By this time production from the OPEC countries had gradually risen once more and the outlook was very gloomy when OPEC met in December, again facing a potential crisis and again unable to find a solution.

The meeting was reconvened a few days later and this time an agreement was reached, leading to an adjustment of prices with good quality crudes coming down in price and the heavier crudes, now back in favour because of changing demand patterns, rising in price. As important as the pricing agreement, however, was the decision to monitor the production levels of member countries, in the hope that this would prevent the widespread cheating and bring production down again, helping to stabilise the market.

This led to a definite tightening in crude-oil supply, which fed through the market to affect products. Stock levels were again low, so there was little chance of substantial destocking in the first quarter of 1985 and prices began to stabilise, albeit with most people looking for a fall later in the year when the effects of another cold winter, this time in Europe, began to wear off.

By this time both Norway and the United Kingdom had decided to abandon the official price structure. Both countries moved towards a spot-related pricing system, which gradually became an actual spot price. The United Kingdom later abolished the British National Oil Corporation and all North Sea producers paid the government royalty direct, also on a spot-related basis. This led to a huge increase in the 'paper' trading of Brent cargoes as companies endeavoured to get the most advantageous price on which to base their royalty payment. This became known as tax spinning.

Paper trading, unlike futures trading, is unregulated (although both the US and UK governments are currently trying to impose some kind of code of practice or regulation) and has no margins or deposits. Like futures trading, however, it is a commitment to buy or sell in the future. The growth of paper trading has made a significant change in the oil market.

Although the OPEC countries maintained official price structures for a while longer than the North Sea, they ceased to have any importance and bore no relation to the prices actually paid by customers.

The next few years saw great volatility in oil prices, with OPEC meetings increasing in frequency but unable to prevent the fall in prices. Observers from other markets claimed that OPEC was bound to fail – no producer cartel has ever managed to survive a falling market. But these predictions eventually proved to be somewhat premature.

One of the most significant events of the next few years was the dramatic slide in prices in early 1986, when the price of West Texas Intermediate (normally seen as the benchmark of international crude-oil prices since the success of the futures contract) fell to $9.75 per barrel.

At this point OPEC saw that its policy of maintaining market share was economically unattractive. It was introduced in the belief that OPEC could dictate the price of crude oil regardless of supply and was further evidence that OPEC was clinging to its old belief that it could control the oil market. Its abandonment suggested that it might be beginning to face up to reality, although it was some time before there was any firm evidence of this in its behaviour or statements.

It tried again to restrict production, with quotas allocated to each country. The next few years saw repeated arguments about the allocation of the quotas, with each country coming up with strong arguments why it should have its quota increased while all agreeing that the overall total should be restricted. Throughout the Iraq–Iran war both countries continued to attend OPEC meetings though for some time only one (not always the same one) was party to the quota arrangements. These quotas were rarely adhered to and prices continued to see-saw.

Most OPEC countries then decided to introduce a new pricing method to try and attract customers back. They began to price oil on a netback basis. The customer bought the oil and then paid the OPEC member the price he received for the products, less an agreed refinery margin. Customers liked the system, because they were guaranteed the refinery margin and refineries had been generally uneconomic for several years. OPEC, however, found that prices fell, because the refiners were not interested in the price they received for their products, but simply in running as much crude as possible to maximise their refinery margins. So this idea was eventually abandoned.

Prices recovered quite strongly during 1987, buoyed by political unrest in the Middle East (including the riots in Mecca) and also by gradually increasing demand in the major industrial countries. But as OPEC increased production, they fell again and it seemed that there was still no real agreement between the OPEC members. Very high US gasoline demand kept prices steady through the summer of 1988, but they began to fall again towards the end of the year.

At this point, however, OPEC finally managed to reach an agreement which, despite early scepticism, appeared to be working. Production quotas were, give or take a bit, adhered to and prices climbed steadily higher through the early part of 1989. It was helped by growing demand. Although OPEC no longer represents the majority of producers, it showed that it could have a greater effect on prices than many had believed, provided that its policies recognised external influences, such as demand changes and economic influences.

The mid-1989 meeting, however, was less cohesive and it is certain that the political infighting in OPEC is far from over. The different objectives of the various countries, ranging from the extremely wealthy, sparsely populated Kuwait to the poorer, high population African and South American countries are not easy to reconcile.

There seems little likelihood that prices will stabilise in the near future, if at all. Fuel substitution and conservation patterns have proved almost impossible to predict. It is clear that the recovery in the major economies in the mid-1980s led to a smaller increase in oil demand than many had expected, though the demand in the developing countries, particularly in South-East Asia rose fast.

OPEC did not control prices in the middle and late 1980s, although its actions had considerable influence because it still represented the major supply of crude oil onto the open markets. Many member countries have begun to invest heavily in downstream operations and are beginning to look a little like the old Seven Sisters, particularly Kuwait with its purchase of Gulf's European refining and distribution system. Late in the 1990s or early in the next century, OPEC will again regain some domination because its members will again be producing most of the crude oil and will also be involved in refining and distribution. But until then, the outlook is likely to remain stormy.

2

Oil refining

The refining process

Once produced, crude oil has to be refined to give the various oil products and it is refining that provides the key to the oil industry. Refinery technology has changed dramatically in the last 20 years, although the basic principles remain virtually the same. There are likely to be more alterations yet, as the industry seeks to come to terms with the changes in demand which are forecast to continue beyond the end of the century.

Crude oil is a complex mixture of hydrocarbons, contaminated with sulphur, metals, salts and other compounds. The principle of refining is the application of heat to crude oil in order to separate the different constituents. A schematic diagram showing the way this is done in a simple hydroskimming refinery is illustrated in Figure 1.1.

Gases

As heat is applied to the crude oil, different constituents boil at different temperatures and can be collected at various points in the distillation column. The temperature decreases as the oil moves up the column, so the compounds with the lowest boiling points, the gases, are collected at the top. The gases are primarily butanes and propanes, with some methane, ethane and other gases. They have a variety of uses, ranging from chemical feedstocks to gasoline additives and bottled gas for domestic cooking and heating. Gases are a small percentage (usually 3–6 per cent) of the total yield from the heated crude oil.

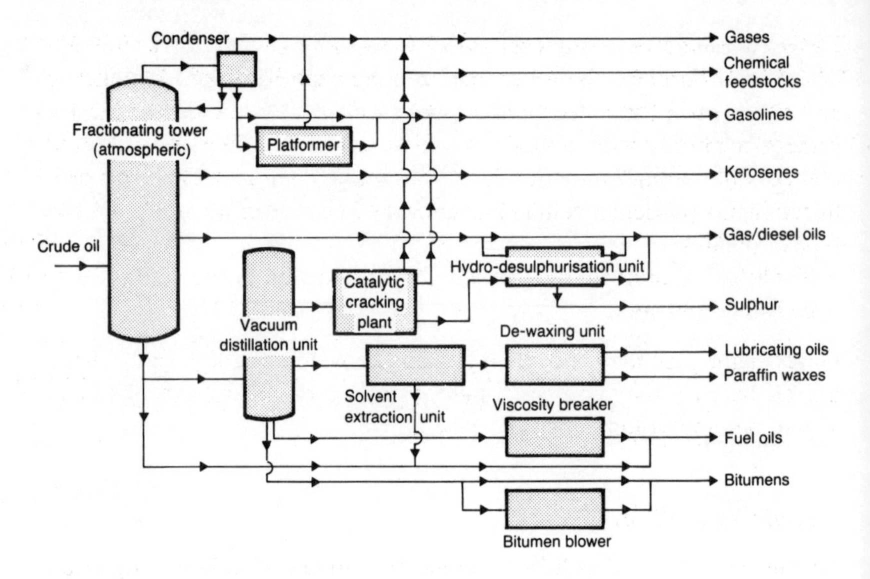

Figure 1.1 A simplified schematic diagram of oil refining.

Light gasolines

Collected at the next stage down from the gases are the light gasolines, which are used for the blending of motor gasoline. These account for only 5–10 per cent of the total yield. Gasoline is a mixture of chemical compounds and is blended to meet local requirements. High-octane gasoline requires a higher proportion of naphthenes (also called cyclo-alkanes) which are made up of six- and five-membered carbon rings, than paraffins, or straight-chain carbon compounds. However, octane numbers can also be boosted by the addition of a variety of other compounds, including lead, toluene and other aromatics and specially developed products such as methyl tertiary butyl ether.

Naphtha

Naphtha is the next product to be collected in the distillation column. This joins the light gasolines in the production of motor spirit. Both fractions are fed to a catalytic reformer, with butane and any other additives, for blending. Naphtha is also a major feedstock for the petro-chemical industry in Europe, though not in the United States, where the

LPGs (liquefied petroleum gases, such as the alkanes) are preferred. There are differences, however, between the naphthas used for blending and those used for petrochemical production. The chemical industry prefers naphthas with a high paraffinic content, while the gasoline blenders like a high naphthenic and aromatic content. (The type produced in any particular refinery is determined by the crude used, not the type of plant.)

Naphthas account for a slightly higher proportion of the whole yield: usually 5–15 per cent.

Gases, light gasolines and naphthas are known as the light end of the barrel, because they consist of lighter hydrocarbons and are lighter in weight per unit-volume than the others.

The middle distillates

At the next stage of collection come the middle distillates, made up primarily of kerosenes and gas oil/heating oil. Kerosenes are mainly used as aviation fuel, though they also have limited applications as domestic burning fuel and for upgrading heavier products. In general, kerosenes are a fairly small fraction: usually 5–10 per cent of the barrel.

Gas oil, the heavier of the middle distillates, accounts for 25–35 per cent of the total. It has two major applications – space heating and diesel fuel. It can also be used by some European companies as a chemical feedstock in place of naphtha. Naphtha is preferred but, when the price differential widens, gas oil can become attractive. Not all European naphtha crackers have this capability to use gas oil, although those that do can have an effect on gas oil prices from time to time. Some smaller industries also use gas oil as a fuel source, but these are few.

Residual fuel oils

Finally, when all the other products have been boiled off, some fuel oils are left. These are used for electricity generation and as industrial fuel and fuel for ships' bunkers (although some ships use marine diesel, equivalent to gas oil).

Increasingly, however, these residual fuel oils are being used to feed upgrading facilities and to produce more light and middle distillate products. Middle distillate production demands a low sulphur content and this will be determined by the crude oil used. Middle East and South American crudes, for example, tend to have a high sulphur content, while North African and North Sea crudes are 'sweet' or low in sulphur. The sulphur concentration in residual fuel is higher than that of the crude it is produced from, because very little of it is removed in the distilling process.

Thermal cracking

There are three methods of upgrading residual fuel: thermal cracking, catalytic cracking and hydrocracking. The first to be introduced was thermal cracking, which works by the application of high temperatures at reduced pressure to increase the degree of distillation. The bonds between some of the carbon atoms are broken by the heat, leading to the formation of lighter products. The use of thermal cracking has been declining in recent years, partly because the newer technologies are more efficient and partly because the gasoline fractions produced do not meet today's engine requirements. As fuel economy becomes more important to the motor industry and the legislation concerning lead content increases, this decline is likely to continue.

Thermal cracking will certainly be used for some years in Visbreakers – the term used to describe the unit which reduces the viscosity of fuel oil by shortening the carbon chains within the molecules, thus making it easier to use and transport. The residual fuel oils can be so thick that they will not move in a pipeline, for example, until they are treated in some way.

In Japan, another form of thermal cracking is popular. Residual oil is heated until all the lighter hydrocarbons have evaporated and coke is left. The upgrading process is cumbersome and atmospherically unpleasant, but the equipment used for it is the cheapest type to install and operate.

Catalytic cracking

Catalytic cracking, currently the most popular form of upgrading residual fuel, works on the same principle but uses a catalyst to enable the treatment to be carried out at a lower temperature. It is not as effective as thermal cracking in increasing the middle distillates, but yields higher-quality gasoline components. Vacuum gas oil is the ideal feedstock for catalytic cracking, but low sulphur fuel oils can now be used successfully. The equipment it requires is more expensive both to install and operate than thermal cracking, but the higher value of the gasoline components makes it a more attractive investment.

Hydrocracking

Chemically, the most efficient upgrading process is hydrocracking. Hydrogen is added to residual oil to increase the proportion of hydrogen atoms to carbon atoms, and to produce more light constituents. This process is more flexible than either of the other two; its major disadvantage is its cost. Hydrogen feedstock for hydrocracking is normally expensive, making the plants costly to operate. The operating costs are around double those of a simple hydroskimming refinery, but gasoline yield is more than three times higher. Most hydrocrackers are currently situated near a cheap form of hydrogen supply, such as a chemical plant.

Changing patterns of demand and consumption

This refining of residual fuel oils in order to upgrade them was needed because of the changes in the demand for different products over the last 20 years. Gasoline demand, as a percentage of all oil product demand, has risen from 12 per cent in 1970 to around 29 per cent in the late 1980s. There has been a significant change in demand for different grades of gasoline within that overall total, most notable being the shift from regular grades to premium and the increasing strength of low lead and unleaded gasolines.

Gas oil demand rose from 29 per cent in 1970 to around 35 per cent over the same period. Diesel cars and increased commercial transport during the economic boom in the West in the mid-1980s contributed to the good demand for gas oil.

Fuel oil has been the product to lose market share over the period, dropping from 40 per cent of all products in 1970 to 18 per cent in the late 1980s. It is the only oil product that can be easily substituted (apart from gas oil's space-heating demand) and many fuel oil consumers are now flexible in their feedstock requirements, capable of taking coal and even gas as an alternative, making price the determining factor in demand.

There have been signs over the last two or three years that relative product demand is stabilising somewhat and forecasts of further falls in fuel oil demand are fewer than they used to be. Refinery technology is changing less dramatically than it was ten years ago, although worldwide capacity is still being upgraded, and price movements have tended to keep demand for all products in roughly constant proportions.

In different areas of the world the demands mentioned above vary slightly. The United States has traditionally been a high consumer of gasoline, mainly because of the large distances covered in 'gas-guzzling' cars. For example, since 1979 the total European demand for gasoline has risen from just over 20 per cent to just under 25 per cent of total consumption, while in the United States it has gone up from 36 per cent to 43 per cent, despite speed restrictions and fuel economy in engine design. In both cases actual consumption has now risen slightly above 1979 levels having fallen around 9 per cent in the early 1980s.

The change in demand has, of course, affected prices since the mid-1970s. The surplus of fuel oils has kept prices low compared with other products. Fuel oil prices have also tended to be largely immune to the volatility of the early 1980s, except during the UK miners' strike in 1984/85, when fuel oil was the strongest product. The low level made most hydroskimming refineries, producing 35–45 per cent of the fuel oil, uneconomic to operate between the late 1970s and mid-1980s. Europe and the United States were suffering from a huge overcapacity in oil refining and it was only those companies with upgrading facilities that could hope to make a profit.

As demand fell sharply in the early 1980s, refinery utilisation dropped and a series of refinery closures was seen, particularly in the major refining areas of the US and Western Europe. The latter saw the largest cutback with closures bringing capacity down to 14 million b.p.d. in 1987 from a peak of 20.4 million in 1978. In the US, refinery capacity peaked at 18.3 million b.p.d. in 1981 and fell to just under 15.2 million in 1987, before beginning to rise again slowly.

For most of the early part of the 1980s refineries made constant losses as refinery closures lagged behind demand falls and products remained in oversupply. Refineries tend to keep operating despite making losses because of the heavy depreciation costs suffered whether the refinery is operational or not. Industry estimates at the time suggested that an actual loss of some $2.50 per barrel was close to breaking-even overall.

For the first half of the decade, many traders believed that the direction of oil prices was being led by products rather than crude.

The drop in demand for oil products was primarily brought about by the recession in the Western world, itself largely due to the rapid increase in the price of oil. But the fall was not only accounted for by the drop in industrial activity. Fuel efficiency and conservation became an economic necessity, leading to developments which, in effect, made much of the demand drop irreversible. High prices also led to dramatic changes in domestic fuel conservation and a switch to alternative fuels for both domestic heating and industrial power generation.

In 1978 oil accounted for 53.3 per cent of total primary energy consumption in the OECD countries, but by 1988 this had fallen to 43.0 per cent. The main beneficiaries have been natural gas, nuclear energy and coal, illustrating one of the main problems with alternative fuel. All of these fuels substitute for fuel oil in power generation and space heating but not for the other major use of oil – transportation. As yet there have been no real alternatives developed for transportation that are both efficient and practical.

Some schemes, such as the Brazilian gasohol project where alcohol produced from sugar is added to gasoline, can be effective in a small area and when the circumstances are right. But in this case a plentiful supply of cheap sugar is essential and as the price of sugar increased in the mid-1980s it became a less attractive option. A similar project was looked at in the US Mid-West using corn as the alcohol feedstock, but this was rejected as economically unfeasible. Because of their use of agricultural products as feedstock, small-scale (in world terms) projects are likely to be initiated from time to time as politically expedient, but there seems little likelihood at the moment that gasohol will really take hold.

Similarly there has been some conversion to LPG as a car fuel. Although this helps spread the transportation load a little further across the barrel, it does not really represent an alternative fuel.

Electric cars also have some way to go before becoming a real alternative, despite the advances made in recent years. They will not become commercially viable until problems such as their limited driving range and long stops for recharging have been overcome.

Gas oil, the middle of the barrel, has seen some substitution, but again for non-transportation use. Natural gas has been the main beneficiary of the change. Gas oil's transportation use remains unchallenged and, indeed, is growing with the increasing move to diesel fuel from gasoline for economic and environmental reasons.

The main challenge to gasoline has come from increased fuel efficiency, and it is difficult to imagine that car manufacturers will be able to continue the scale of the improvements made in the last ten years. This aspect of demand has also shown itself to be one of the most price-sensitive areas of oil consumption. As gasoline prices rose in the early 1980s, the US saw a move away from the 'gas-guzzling' cars of the 1960s and 1970s towards smaller, fuel-efficient cars, but low prices in the second half of the decade saw a swing back towards less efficient cars.

This pattern of consumption, with increasing demand for the top of the barrel, has led to increasingly severe oil refining, where crude oil is treated at higher temperatures and with a variety of catalysts to convert a higher proportion of the crude to lighter products. This is inevitably more expensive than older methods and has required massive investment from the refining industry.

The refining industry

The proportions of the different products made in a refinery are heavily dependent on the type of crude oil processed. The heavier (literally) crude oils of the Middle East and South America give rise to high volumes of fuel oil but little gasoline, whereas the lighter crude oils of the North Sea, North Africa and the United States produce relatively small amounts of fuel oil, with a low sulphur content which is therefore able to be cracked more easily. (In most modern refineries the heavier products, after straight-run refining, are passed on to a secondary cracking process.)

This has made light crudes very much more attractive than heavier ones through most of the 1980s. This tendency is unlikely to change, for the reasons given earlier, except when occasional external factors, such as the UK miners' strike in 1984/85, alter the balance. The increase in fuel oil cracking has led to some respite for fuel oil, but this is likely to alter again as cracking efficiency increases and technology improves.

Until the early 1960s the major oil companies owned and ran virtually all of the world's oil refineries, which were largely located in areas convenient to the markets for their products. Thus there are, for example, several refineries in the Amsterdam–Rotterdam–Antwerp (or ARA) area of northern Europe, with its ready access to the Rhine and thence to West Germany, Switzerland and the inland areas of Belgium and The Netherlands. Similarly, in the United States there are a number of refineries on the Gulf coast of Texas, where crude oil is produced and imported and access to the inland US markets is easy, though in this case the ease of access has come about more by design.

During the late 1960s and early 1970s, the increasing nationalisation of crude-oil production broke the oil companies' monopoly over the passage of oil from well to consumer. As a result, the oil spot market flourished. With more crude oil available other companies were able to get involved in refining and this led to the growth of independent refineries. They tended to be built where access to crude-oil supply was easy, hence the number of such refineries located on the Italian coast. The independent oil refiners bought crude from the spot markets, processed it and then sold the products back on to the spot market, obviating the need for their own integrated distribution systems.

German legislation, passed in the 1960s to open up the distribution of oil products and reduce the monopoly of the oil majors, also contributed to the growth of independent refining: here was at least one major consumer with an 'open' market. Gradually, independent marketing companies appeared in other countries too, buying oil products from the spot market and retailing them through small distribution systems.

It was the growth of the refinery industry on the Italian coast, close to the supply of North and West African crudes, which led to the formation of the secondary European spot market in the Mediterranean. Products dealt in this market are largely consumed by the Mediterranean countries themselves, but when a supply shortage exists in the northern part of the continent, the Mediterranean market can be tapped. Similarly, cargoes of gas oil and gasoline are shipped to the United States whenever prices are right.

These independent refineries prospered in the mid-1970s as oil demand grew and prices were steady. Although they had the effect of reducing the profit margin of the majors, there were still profits to be made in oil refining until after the 1979 price boom. The subsequent drop in demand and prices for all oil products made refining uneconomic for everybody, big or small, but it put the independent refiners in particular difficulty because they lacked the majors' integrated supply systems and stronger financial position.

Oil-refinery economics can be talked about in general terms, though each individual refinery and type of crude oil can generate entirely different figures. In order to run a profitable refinery the selling price of the products must be greater than the combined cost of crude oil, of running the refinery, capital depreciation and transportation and distribution costs. Although this seems self-evident, it was rare in the early 1980s for such profits to be made. And the marginal cost of

Table 2.1 World oil consumption (million barrels per day), 1975–88

	1975	1976	1977	1978	1979	1980
United States	15,875	16,980	17,925	18,255	17,910	16,460
Canada	1,735	1,790	1,810	1,835	1,915	1,855
EEC*	11,545	12,330	12,085	12,515	12,805	11,850
Other Western Europe	1,720	1,870	1,860	1,850	1,875	1,815
Japan	5,020	5,190	5,350	5,420	5,485	4,925
Australasia	695	725	750	750	755	720
Latin America	3,520	3,570	3,820	3,975	4,230	4,400
Middle East	1,320	1,475	1,565	1,620	1,495	1,625
Africa	1,050	1,155	1,205	1,290	1,360	1,450
South Asia	610	665	705	760	775	825
South-East Asia	1,535	1,675	1,805	2,015	2,290	2,325
Total free world	44,575	47,380	48,830	50,235	51,280	48,605
China	1,350	1,530	1,630	1,705	1,835	1,765
USSR	7,520	7,780	8,125	8,480	8,640	9,065
Total centrally planned economies (CPEs)	11,105	11,740	12,350	12,860	13,230	13,560
Total world	55,730	59,165	61,230	63,145	64,124	61,820

*As at 1983
Source: BP

refining, by ‘forgetting’ capital depreciation and some running costs (such as a labour force which must be paid regardless of the operational capacity), is very much less than the real cost. Thus, whenever the refining of crude oil has been economic on a marginal-cost basis, the utilisation of refinery capacity has increased and, since the market for no oil product has seen sufficient long-term strength to push refineries into real profit, traders tend to talk of refinery profitability whenever the gross-product worth of a barrel of products exceeds the cost of the crude by more than 50 cents or so.

The product slates of different crude oils can be varied within a refinery by adjusting the severity of the processing and by upgrading various proportions of the products. With this sort of variability in the mixture of end products it can be difficult to estimate refinery profitability. But average yields of different crudes are published from time to time by the oil journals and can be used to give an approximate estimate of profitability. The precise profit of any individual refinery is less important to the trader (within reason) than the trend and, by following a consistently calculated estimate, some idea of profitability can be obtained.

1981	1982	1983	1984	1985	1986	1987	1988
15,550	14,765	14,745	15,175	15,170	15,670	15,955	16,420
1,760	1,565	1,475	1,425	1,490	1,475	1,540	1,615
11,095	10,570	10,365	10,455	10,260	10,645	10,615	10,745
1,725	1,650	1,605	1,630	1,640	1,735	1,730	1,735
4,695	4,395	4,390	4,565	4,380	4,435	4,435	4,760
710	695	670	695	670	695	730	745
4,435	4,540	4,450	4,420	4,425	4,530	4,740	4,825
1,780	2,050	2,285	2,155	2,150	2,150	2,705	2,815
1,525	1,610	1,630	1,630	1,700	1,695	1,775	1,795
890	935	950	1,030	1,160	1,210	1,260	1,370
2,375	2,300	2,310	2,315	2,275	2,450	2,545	2,890
46,835	45,165	44,965	45,835	45,800	47,145	48,075	49,765
1,705	1,660	1,705	1,735	1,745	1,820	1,925	2,035
9,085	9,030	8,905	8,890	8,885	8,930	8,920	8,890
13,365	13,185	13,030	13,040	13,100	13,310	13,380	13,480
60,200	58,350	57,995	58,875	58,900	60,455	61,455	63,245

Refining was generally uneconomic through the first half of the 1980s with odd bouts of real or marginal profitability, usually a result of cold weather or some other short-term increase in demand. A sharp cutback in the refining capacity of the Western world continued until the middle of the decade, when capacity began to increase again largely because of new refining areas such as the Middle East and Pacific Rim countries entering the field, although the US and European capacity also began to increase slowly.

Despite the cutbacks, utilisation rates also fell, with Western Europe and the United States reaching lows in 1981 of 56 and 68 per cent, respectively. Later in the decade, utilisation rates rose again, led by the gradual increase in demand. The United States saw the first real benefits with high gasoline demand in the summer of 1988 keeping refineries running at around 85 per cent through most of the year. Table 2.1 gives a breakdown of world oil consumption figures between 1975 and 1988.

Another major factor affecting the Western refining industry has been the increase in refining from the OPEC countries and other areas, particularly South and South-East Asia. In the Middle East, for example, refining capacity rose from 3.6 million b.p.d. in 1980 to 4.3 million b.p.d. in 1988 and in Indonesia alone from 0.5 million to 0.8 million.

Most OPEC countries are becoming increasingly involved in the downstream side of the industry either by building their own refineries, as in the Middle East, or by signing long-term supply contracts that involve some transfer of equity with importing nations, such as deals done by Venezuela and Mexico.

The long-term effects of these moves will probably take a number of years to make themselves felt, but it does seem possible that OPEC could begin to exert similar powers to those once held by the majors in some areas. Such control will not really be possible until the oil supply balance tilts back in OPEC's favour, probably some time in the late 1990s.

3
The markets

The spot market

The history of crude oil outlined in the first chapter cannot be considered on its own in an attempt to understand today's oil markets. During most of the time markets have existed outside the major oil companies, crude oil has been the dominating influence on the price of all oil products. The only real exception to that came in the early 1980s when oversupply led to an increasing influence for oil products and it was really the lack of demand for these that led to the sharp falls in prices. Towards the end of the decade as crude supply was restricted, crude oil moved back towards its dominating position, though the seemingly insatiable US gasoline demand seemed unlikely to be ignored.

Prior to the 1960s and 1970s, virtually all of the world's oil-refining capacity was in the hands of the oil majors. But gradually the independent companies began to set up refineries. Then came the independent refineries, set up for the sole purpose of buying crude on the growing spot market, processing it and selling the products, sometimes under long-term contracts but more usually back on to the spot market. It was these last that really led to the changes. Whenever refining could be done at a profit these companies stepped in, but whenever refinery economics became too gloomy they withdrew from the market altogether. The majority of these refineries were set up in the Amsterdam–Rotterdam–Antwerp area, the Mediterranean coast of Italy, the Gulf of Mexico, the Caribbean and, latterly, Singapore.

These independent refineries were both a natural result of the oil spot markets and a major factor in their development. The spot markets had begun to emerge as more participants appeared in the market and the supply chain from oil well to consumer was no longer in the hands of single entities. They evolved from the need for a balancing mechanism to handle excess supply and demand, and this in turn enabled the independent refineries to use the spot market as a source of supply and a market for products.

Until the early 1980s the oil majors used the markets only rarely and admitted they might be useful even less often. But as refining became uneconomic and long-term supply contracts were abandoned, the majors were forced to turn to the markets more and more. BP, the first of the former Seven Sisters wholeheartedly to adopt a trading mentality and the only one to be net short of crude oil at the time, announced in early 1983 that it was buying more than 50 per cent of its crude oil from the spot market.

The spot or cash markets are based on five major centres and a number of other smaller ones, although the actual trading often takes place many miles from the nominal centre. The five major spot markets today are: North-West Europe (loosely based on the ARA area though the cargo market works out of London); the Mediterranean, based on Italy's west coast; the Gulf of Mexico, out of Houston; the Caribbean; and Singapore, the most recent addition to the list and the fastest growing.

The North-West European market

The North-West European market is the larger of the two on the Continent, covering, as it does, four of the five major European consumers – West Germany, the United Kingdom, The Netherlands and most of France.

Spot market trading in Europe began in the 1950s, when it was a very small and insignificant part of the industry, trading very small volumes and being largely ignored by the oil companies, which still had marketing their own way. In terms of volumes traded it is still fairly small, representing about 5 per cent of the total European product trade, but it is by no means insignificant. Indeed, at times its significance moves out of all proportion to its size.

The volumes handled are virtually impossible to determine, because a cargo frequently changes hands several times before arriving with an end-user and there is no record of trading apart from those held by the individual traders. Approximately 50 million tonnes are traded each year, but only about half of this actually changes hands. There is no official record of prices. Although the deals done are generally public knowledge (with the names of both buyer and seller, and the prices, talked about) it is all done by word of mouth.

There are a number of price reporting systems assessing the deals done on a daily basis and publishing them on screen or telex services. They all rely on using a team of people to telephone around a number of traders to find out what deals have been done and at what price. Although these services were widely believed to be inaccurate and easy to fix in their early days, the increasing number and professionalism of them has improved their accuracy and acceptability. They are now used as a basis for settlement on the International Petroleum Exchange (IPE) crude-oil contract and are being looked at by the IPE and other exchanges for new contracts where physical delivery problems appear otherwise insurmountable. The EEC did try to run a monitor on the spot market, but it was resisted by the traders, who mostly refused to participate, and has now been abandoned.

The USSR is the major source of supply for the North-West European market, primarily of gas oil, which accounts for around half of all the spot market trading. Gasoline also comes from a number of East European countries, and is frequently blended and then shipped out to the United States.

The USSR is the largest oil producer in the world, with an output of 12.7 m.b.d. or some 21 per cent of total world production. It has agreements with East European countries for some of this production, but more than half of the total is exported as either crude or products. Apart from sales of gold, oil provides the USSR with its best source of foreign currency and exports are therefore stepped up when dollars are required. Traditionally, little gas oil was exported during the winter months, when domestic demand was at its height, but in 1981/82 and the following year bad harvests necessitated a higher flow of dollars, and exports continued throughout the year, putting further pressure on already weak oil markets.

Another source of supply is the independent refineries around Europe, either in the ARA (Amsterdam–Rotterdam–Antwerp) area or elsewhere. (Although the market is often thought of as being in The Netherlands, this is not really true.) In general, independent refineries operate only when the economics look attractive.

In the late 1970s, several large trading companies bought refineries, but as refinery economics worsened in the early 1980s they became a distinctly unattractive asset. On the whole, the independents were less technologically advanced than the majors. A number of the independent refineries were mothballed or closed altogether in the early/mid-1980s. Renewed interest was not apparent until recently.

The overcapacity in the first half of the decade led to drastic cuts in capacity and major upgrading, as discussed in more detail in the previous chapter. Total world capacity then began to return to some sort of balance, though there was still a dislocation between the refinery sites and the areas of consumption. And with capacity likely to build further in OPEC countries and South and South-East Asia this dislocation may become even more exaggerated.

As well as operating for themselves, independent refiners will also refine crude for other people under a processing deal. They will usually charge either a per barrel fee for refining or they will take a proportion of the products made. Opportunities for attractive processing deals are infrequent and usually depend on access to cheap crude. This is becoming increasingly unlikely as crude oil producers, other than the major oil companies, move towards tying up crude supply long term with adjustable price contracts or try to take a long-term stake in a refinery in return for supply.

This has the benefit, however, of making refining on their own account more attractive for the owners of the refineries. Towards the end of the 1980s, refiners were being actively courted by some of the OPEC countries.

Another important factor in the European market is independent storage. These tanks abound in the ARA area and the exact level of stocks is known to no one but the operator. Most of these independent tanks are owned by the traders. Much of the oil traded on the spot market (and all that delivered on to the IPE) is stored in these tanks. There are similar installations in the other spot market areas, but in the United States, for example, demand and production figures for stock are published weekly, so a closer eye can be kept on changes in demand. In the ARA area, stock levels are assessed by traders flying over the area and watching the level of the floating tops on the tanks!

In Europe, there have been several attempts to monitor stocks in a similar way to the United States, but they have mostly failed because there is no legal requirement to report. The EEC has tried on several occasions to institute a monitoring system and the latest effort has enjoyed more success than most, but has still not become such an institution as the US American Petroleum Institute (API) stocks. There are a number of consultancy firms offering monitoring services with some degree of success.

There are two parts to the North-West European market – barges and cargoes. The barge market trades in 1,000–2,000 tonne parcels of oil products largely for movement down the Rhine into West Germany and Switzerland. The term also covers the movement of small quantities of oil into the United Kingdom and France. The source of supply for the barge market is primarily the majors' oil refineries on the North-West European coast.

There are a large number of barge traders, many of whom are based in Rotterdam, who trade the barges speculatively as well as moving them from refiner to distributor or direct to end-user. Although the speculative element of barge trading has declined since the advent of oil futures trading, because of the lack of operational difficulties in the paper market, it is still a major part of the industry, resulting in a much larger apparent volume of trade than can be accounted for in barge movements. None the less, the overall level of activity is somewhat less than it was in 1979/80, with falling demand a major factor. Other important causes are a series of periods with little price movement for some weeks at a time and a readjustment amongst traders after the dramatic price movements in earlier years.

Barges are normally traded on a free-on-board (f.o.b.) basis in the ARA area.

The other major sector of the oil-product spot market is the cargo market. Although this is centred in the places mentioned earlier, it is a more international market, with cargoes frequently moving from one market to another. Each parcel in the cargo markets is normally between 18,000 and 30,000 tonnes, usually priced on a c.i.f. (cost, insurance and freight) basis in North-West Europe, but often on an f.o.b. basis elsewhere.

All products are traded on the cargo market, but in Europe gas oil accounts for around half of the market, largely because of the high level of speculative trading. In particular there is an active market, known as 'Russian Roulette', in which cargoes of Russian gas oil change hands as many as thirty or forty times. This has grown in recent years and has led to the formation of two almost entirely different gas-oil spot markets: one in non-EEC (usually Soviet) gas oil and the other in EEC-qualified material, which is more supply–demand based and very much less speculative. Although the distinction between the two is also apparent on the barge market, where it is more simply a difference in designation and area of delivery, it has less impact than on cargoes.

In the United States gasoline is the largest-volume product traded, with heating oil second in line.

The historical basis for the high proportion of gas-oil trading in Europe is the independent distribution system for home heating oil (gas oil covers both home heating oil and diesel fuel). Domestic heating oil can be distributed in fairly small volumes and does not require the same infrastructure as, for example, gasoline. In many cases former coal distributors went into oil when oil began to displace coal as a home heating fuel.

In West Germany legislation was introduced in the 1970s, encouraging the setting up of independent marketing companies, in an attempt to reduce the monopoly of the Seven Sisters. These distributors were able to look around for the best source of supply, be it the major oil companies or the independent refiners now putting product on to the growing spot market.

The 'Russian Roulette' trading of gas-oil cargoes from the USSR has also contributed to the large volume of gas oil traded, but by its very nature it is difficult to quantify.

Another reason that gas-oil trading is more active is that gasoline specifications, particularly on lead content, vary widely across Europe, thus constricting the market somewhat.

In the United States the situation is very different, with gasoline and heating oil (like gas oil the term is also used to cover diesel fuel) accounting for very similar proportions of the total. This is not only because of the, until recently, country-wide gasoline specifications but also because gasoline and heating oil account for a similar proportion of total oil demand. Gasoline accounts for around 42 per cent of the US oil-products market, whereas in Europe it comes second to the middle distillates (gas oil plus kerosenes) with only 25 per cent of the market.

The companies active on the spot market can be divided into two very distinct categories: brokers and traders. Brokers do not take a position on the market, they simply act as intermediary between a buyer and a seller, taking payment in the form of a commission. Payment for the deal passes directly from buyer to seller and the broker's involvement ends when the deal is agreed. They are frequently used as a means of keeping the identity of the buyer and seller secret until the deal is arranged – particularly useful when, for example, a major oil company wishes to use the market without word spreading. In the United States they also provide a means of complying with anti-trust legislation for the major oil companies.

The traders, on the other hand, take positions on the market, buying and selling speculatively rather than just to offset demand and supply. They expose themselves to large financial risk in the hope of equally large reward. Consequently, every time the price of oil makes a major move, some companies tend to disappear. The oil majors mostly now engage in trading in this way, not simply buying in shortfalls in product and selling surpluses.

Brent Blend cargoes of crude oil are traded in a similar way to the Russian gas-oil cargoes, with one cargo changing hands many times before an actual delivery is taken. Brent, in addition to being the largest-volume crude in the North Sea, is the largest-volume, non-OPEC crude on the free market (though the United States and USSR produce more) and has become a highly speculative market. Each cargo of 500,000 barrels may be traded many times before it is finally collected from Sullom Voe, the loading terminal for crude from the Brent and some other fields. There are around ten or eleven major traders in this market, along with about twenty smaller ones. The major oil companies and refiners also participate actively in this market. It is not unusual to find one trader or oil company several times in one chain, as a cargo moves through as many as thirty or more links. Crude oil is, however, a truly international market and does not fit into any of the specified geographical areas.

There is now a separate category of trader, known as the Wall Street refiner. A number of the US investment banks set up trading arms to deal in oil in much the same way as they deal in other financial instruments. They are assuming the risk for a number of companies involved in the oil markets and laying off those risks in the physical or futures markets in much the same way as an insurance company. Several of them also take substantial outright positions on their own account. They have only been in existence since early 1987 and made an immediate impact on the oil industry worldwide.

Their customers come from right across the industry from producers to consumers and they offer more flexibility than the paper or futures markets are usually able to, tailoring options and other instruments exactly to match the customer's requirements. They will then study all the available ways of offsetting the resulting risk and lay it off wherever seems most appropriate. Their charges will normally be built into the cost of the instrument offered.

The Mediterranean market

The smaller of the two European spot markets, the Mediterranean, is supplied primarily by local refineries, particularly the independents situated on the West Italian coast and the islands. There is some supply from the USSR via the Black Sea. The Mediterranean market tends, however, to be less volatile than its northern counterpart.

The Caribbean market

This is the smallest of the recognised oil spot markets but has an important role to play in balancing supply, particularly on the US and European markets. Crude oil is produced and refined in the area and normally shipped to the US market, though gas oil and fuel oil sometimes find their way over to Europe. The market does not trade actively.

The Singapore market

The fastest growing spot market is that in Singapore. It is the newest of the main spot markets, having been trading for only about ten years, but has established itself as the centre for trading in South and South-East Asia. This area is primarily served by the developing local refining industry and the Arabian Gulf refiners.

The high demand for light products in the Western world has meant that the heavier crude oils produced in the Middle East tend to go East, though increasing production in the region and the slight swing back to heavier crudes forced on the industrial West by the lack of sufficient supply of sweet crudes has begun to change this.

The Singapore spot market has flourished, strongly supported by the government, and a futures contract in heavy fuel oil was introduced in early 1989. Fuel oil and naphtha have been the major products traded in the area, naphtha primarily because of the Japanese import requirements.

Singapore has become the focus of attention for countries as far apart as India and Australia and now enjoys all the infrastructure of the older markets.

The US market

The United States is the second-largest crude-oil producer in the world, with a daily output of some 9.5 million barrels. The remaining seven million barrels per day (approximately) that it requires are met from imports from all over the world, but primarily South America, the United Kingdom and Nigeria.

On the Gulf Coast and in some other centres, including New York and Southern California, there are active spot markets similar to those in Europe, where parcels of product and crude are sold from trader to trader. But the character of the US markets is made very different by the pipeline systems which exist to transport crude oil and products around the country. This makes the parcel-size very much more flexible than the shiploads traded in Europe (and goes some way towards explaining the runaway success of the crude-oil futures contract). Elsewhere, for example, crude oil tends to be traded in 400,000–500,000 barrel parcels, but in the United States volumes as small as 10,000 barrels may be traded, though larger volumes are more frequent. This has led to a very much more active crude-oil spot market, with a much larger number of participants than in Europe, where the financial commitment is so great as to deter all but the largest companies and traders.

Price differentials have traditionally existed between Europe and the United States for a number of reasons, one of the most important being the cheap domestic crude. Although only the USSR produces more crude oil, the US government forbids crude oil exports and the country has never been a major force as a producer in the oil markets. Domestically produced crude has never had to compete on the international market and the price remained unnaturally low, aided by import restrictions on oil products until 1980/81. Even after decontrol, the lower prices continued until after the spectacular price falls of early 1983, since when prices have tended to move more or less in line with international crudes of similar quality.

During the price recovery, oil-product demand in the United States was increasing for the first time for several years as the economic recovery got underway. As this recovery was not being seen in Europe, where oil demand was still falling, there was a considerable movement of almost all oil products, even the unattractive fuel oil, across the Atlantic, keeping the European spot market very much steadier than local conditions warranted. The interaction between the two largest spot markets (taking the United States as one market) is likely to continue, in the absence of any restriction on movement as the market seeks to readjust to the changing level of demand for different products; this topic is discussed in Chapter 2.

The increasing level of crude imports to the United States has been of concern to the government for some time and whenever crude oil prices have fallen below around $15 per barrel for any length of time the question of an import tax on crude oil is raised. It is unlikely that domestic production can be increased significantly, except in Alaska where environmental issues have long been an important factor in assessing development potential, even before the major oil spill in early 1989. There is some production shut in in Texas and elsewhere since the low prices, but this would not be of great long-term significance.

The futures markets

The oil futures markets were set up to enable traders to offset some of the risks they take; by hedging their position, or taking the opposite futures position to that which they hold on the physical market. Thus a trader who has bought a cargo of gas oil would sell futures to protect himself against a fall in price before he can sell his cargo.

Commodity futures markets developed in the late eighteenth and early nineteenth centuries as trade grew first nationally and then internationally. They developed from the corn exchanges seen in almost every town of any size, where merchants, producers and consumers used to gather to trade. As the time lag between growth and actual sale grew, because of the distances covered in the transit of products, it became necessary to hold stocks and anticipate future supply and demand. Markets began to be affected by non-local factors and prices became more erratic. These developments were particularly noticeable in international commodities such as cotton, sugar, cocoa and coffee where there were several weeks or months between harvest and sale.

Many people date futures trading from the American Civil War when English cotton mills bought American cotton before it had been shipped; though there is some evidence that futures trading existed in some ancient civilisations.

Although the origins of modern futures trading were in England, the largest centre of commodity trading is now Chicago, where there are two large commodity exchanges trading contracts in a wide range of agricultural products, financial instruments and metals. Contracts range from orange juice, meat products and grains through precious and other metals to foreign currencies, bonds and stock market indices.

There has been rapid growth in futures trading since the mid-1970s, with a large number of contracts opening up, particularly in America. Many of these contracts, like the energy futures in Chicago, the first fuel oil in New York and the first crude oil in London, never really get off the ground, but a large number do become successful, with some of the financial instruments in Chicago recording the trade of more than 200,000 contracts in a single day.

It will take a long time for oil futures trading to become as fully integrated with the oil industry as, say, sugar futures are with the sugar industry. But a large number of those involved in the industry over the last ten years would admit that they never thought the markets would become the influence they have and, more importantly, would become as useful as they have.

As the oil markets generally become increasingly international and the reporting systems disseminate information instantly, the opportunities for older trading techniques have diminished and new methods are having to be found. Futures trading has opened up a number of new possibilities, many of which are being used increasingly by an industry used to adapting itself quickly to outside changes. Perhaps the most useful function the markets can perform long term is the separation between price and supply, and it is here that the greatest growth in futures business has come in the last two or three years.

4

The futures contracts

In late 1989 there were nine energy futures and five energy options contracts trading in three locations, with several others under discussion and some certain to begin trading. The most successful is the New York Mercantile Exchange's West Texas Intermediate crude-oil contract, which averaged around 75,000 contracts (75 million barrels) per day during 1988 and has traded in excess of 135,000 contracts in a day.

The NYMEX also operates the second and third ranked contracts – heating oil and unleaded gasoline. Heating oil was the first successful oil futures contract, introduced in November 1978. The other futures contracts traded are propane and fuel oil in New York, gas oil, fuel oil and Brent crude on London's International Petroleum Exchange and fuel oil on the Singapore International Monetary Exchange. Options are traded on all these contracts except propane and fuel oil.

Propane

The propane contract is the oldest currently trading, although it has transferred from the New York Cotton Exchange, where it started life in 1971, to the NYMEX. It is not a successful contract, despite its longevity, and still only trades a few contracts a day.

Between its inception and the opening of the heating-oil contract in 1978, there were a number of attempts to introduce energy futures trading elsewhere, in both Europe and the United States. There were two reasons why these were unsuccessful: first, and most importantly, was that the time was wrong; secondly, the siting of the markets was wrong. Timing is essential for all new futures markets. Futures markets depend on volatility in prices and the 1970s saw, more or less, a steady rise in prices, at least until the Iranian revolution. Even the heating-oil contract had a very quiet start because prices were effectively moving in one direction.

The siting of markets in major financial centres was also important because all futures markets require a financial infrastructure to support the clearing mechanism. This was missing from most of the early attempts at introducing new contracts.

NYMEX No. 2 heating-oil contract

In 1978 the NYMEX introduced its No. 2 heating-oil contract and a No. 6 fuel-oil contract. Interest from both the oil industry and the essential speculative element of the market was slow to develop initially and the fuel-oil contract failed to progress. After a time trading stopped altogether. Heating oil, on the other hand, very slowly began to attract interest and the violent price rises of 1979 and 1980 enabled the market to establish itself.

As a product, heating oil satisfied the major criteria necessary for an active contract – it was heavily traded on the free market, relatively easy to specify, store and transport and, as a result, to deliver. Although physical delivery is not, and should not be, the main function of a futures market, every contract must be supported by a sound, realistic delivery procedure to gain the confidence of the industry and to ensure a close price correlation between the 'paper' futures contract and the 'wet' physical oil market. Nevertheless, in its first year or so, the heating-oil contract failed to capture the imagination of the oil industry in New York and the US coast of the Gulf of Mexico. The market traded very low volumes, sometimes only one or two contracts per day, and was not making the progress that had been hoped for. Gradually, however, the exchange's marketing and education programme, at the time the most ambitious and deliberate campaign ever mounted by a commodity exchange, began to bear fruit. Interest in the market began to increase and, as the traders tested the market, liquidity began to improve and reach sustainable levels, attracting still more interest. By the end of 1984, the market was regularly trading more than 20,000 lots per day and on occasion has reached 35,000 lots. The open interest averages around the 65,000 mark. (Open interest, which represents the number of lots which remain uncovered on any day and would therefore have to be delivered if the market were to cease trading, is often taken as a better guide to the liquidity of a market than its daily turnover.)

The slow start was inevitable in retrospect. The oil industry was positively antagonistic to the concept of futures trading, entirely new to almost everyone in the trade. This antagonism was to be repeated in Europe a few years later when the London market opened. The list of advantages attached to futures trading is usually headed by the opportunity of 'hedging', the laying off of risk and locking in a profit in case the market turns against you. This idea was anathema to an industry which had made vast sums of money by taking on the risks they were now being told to avoid.

Probably the main reason for the collective change of mind which has been seen is the two-directional volatility seen in oil prices since 1980. Before, oil prices had only ever been stable or moved sharply upwards – weakening levels were almost always short-lived and slight. But this has certainly not been true since 1980.

IPE gas-oil contract

After the New York heating-oil contract, the next to be introduced was the International Petroleum Exchange's gas-oil contract in London. Gas oil is the European name for heating oil and, although there are some specification differences between the two contracts, the product is essentially the same. When the contract was introduced, in April 1981, the IPE began a similar, though smaller scale, marketing campaign to that carried out by NYMEX three years earlier. A small proportion of the European industry had used the heating-oil contract, but essentially the IPE faced the same problems NYMEX had had. US import controls had meant that there was not necessarily any correlation between US and European prices, so the traders that had tried out the New York market had tended to treat it as a bit of a game.

The London gas-oil contract had a steady start, trading some 800 lots per day during 1981 and gradually increasing until, during the uncertainty in the winter of 1988/89, the market averaged over 7,000 lots per day, with over 15,000 lots traded on the busiest days.

Like the NYMEX in New York, the IPE adopted an active marketing policy. It was the first London commodity market to do this and the idea was treated somewhat sceptically by its soft commodity and metals equivalents. The success of the approach won over the other markets, however, and the approach was adopted more aggressively, and even more successfully, by the London International Financial Futures Exchange a few months later.

NYMEX leaded-gasoline contract, heating oil and gasoline (Gulf Coast delivery contracts)

Having established a solid trading base, both exchanges were well placed to introduce further contracts. The NYMEX was the first to take the plunge, introducing a leaded regular gasoline contract in late 1982. The contract was reasonably successful, growing quite well although it remained less active than heating oil until it ceased trading (having been replaced with an unleaded gasoline contract) in 1986. The unleaded gasoline contract is fully established and trades a greater volume than heating oil much of the time. In October 1989, NYMEX introduced a 1 per cent sulphur fuel-oil contract. At the time of writing it was far too early to make a judgement about its likely future.

NYMEX also introduced heating-oil and gasoline contracts with US Gulf Coast delivery, but these failed to attract any business at all. In the case of heating oil particularly this was largely because the attractions of a liquid contract outweighed the locational disadvantages of the contract. It has always proved extremely difficult to introduce a second successful contract in any commodity within one time zone.

NYMEX crude-oil contract and CBT crude-oil contract

Six months after the gasoline contract, four new markets were opened, one on the NYMEX and three in Chicago. New York opened a crude-oil contract, based on West Texas Intermediate; and the Chicago Board of Trade (CBT) brought in heating oil, unleaded regular gasoline and crude oil, based on Light Louisiana Sweet. Of these, only the NYMEX crude contract had any degree of success and the three contracts in Chicago ceased trading.

Despite the success elsewhere of the oil-product futures contracts, the CBT was unable to attract interest from the oil industry, and without that the speculators lost interest. There were several reasons why the trade was inactive. Firstly, the industry had established good contacts and relationships with the commodity brokerage houses in New York and saw no reason to change when the Chicago market had nothing different to offer. This attitude was compounded by the CBT's marketing approach, which did not stand comparison to the aggressive New York version. It appeared that the CBT preferred to rely on Chicago's reputation as the world's most active centre of commodity trading, and wait for the industry to come to it rather than sell its services to the industry. Finally, the delivery procedure chosen for the heating-oil and gasoline contracts was at best cumbersome and certainly far more complex than that in New York, thereby failing to meet the requirements for a delivery procedure closely allied to physical market practice.

The New York crude-oil contract, on the other hand, quickly became the most successful futures contract ever introduced (if you use open interest as the measure of success), attracting enormous interest from trade and speculator alike and rapidly becoming a focal point for the entire oil industry, even those not actually using the exchange. The success of the contract did not dispel criticism from some parts of the industry, but the volume of trade and the quality of the participants soon made it impossible to ignore, even for the most die-hard industry conservatives.

Within eighteen months all but two of the original Seven Sisters (now Six Sisters, since the Chevron and Gulf merger) were using the exchange to some extent, and its influence was to be felt throughout the world. Virtually all US and European oil refiners and traders now use the exchange, and Asian and Australasian traders, too, are becoming increasingly involved. The market offers protection for traders in crudes other than West Texas Intermediate by using the differential between types of crude.

The 'exchange for physicals' method of delivery, whereby a long and a short can mutually agree to deliver any crude at any place, with a differential price against the WTI futures if necessary, has given a major boost to the contract. Almost all the deliveries against the crude-oil futures contract are made using this procedure (it is impossible to quantify the amount because the growth of the market is such that deliveries increase every month).

IPE crude-oil contract

The next energy contract to be introduced was the International Petroleum Exchange's first crude-oil contract, based on the primary North Sea crude, Brent Blend. This started trading in November 1983 but quickly faded into obscurity.

One major problem faced by the IPE was delivery of crude oil. Under the rules of trading, it must be possible to deliver one lot of a commodity traded on the futures market. In the United States, with its pipeline systems and major terminals such as Cushing in Oklahoma (chosen by New York) and St James in Louisiana (by Chicago), the delivery procedure could be fairly simple. It is possible to deliver relatively small (5,000 barrels) quantities of crude oil. In the United States, anyone delivering less than five lots of crude oil must do so in a tank storage installation. But in Europe, even 5,000 barrels is too small for an oil terminal to handle – the whole system is geared to quantities of 400,000 barrels upwards.

Thus, the IPE crude contract wilted. But the European crude-oil trade, although using the NYMEX contract actively, has none the less felt the need for a crude futures contract more closely allied to their own business and pressure was exerted on the IPE to find some way of developing a new Brent futures contract as soon as possible. This pressure was increased during the winter of 1984/85 when the differential between WTI and Brent moved more than the absolute price of crude oil, making WTI unacceptable as a hedging vehicle to most traders, and again when BNOC was disbanded.

A number of ways of developing a new market were studied. The most attractive was the establishing of a crude-oil price index, based on published price data for spot deals on the Brent market, which would be used as a basis for cash settlement. Thus, when a delivery month expires, and delivery would normally take place, physical oil does not change hands, but the current market value of the oil does, enabling the buyer and seller to achieve the same financial position they would have been in with the oil. Such an index was introduced in May 1985.

The oil industry's reaction was cautiously approving and a contract based on cash settlement began trading in November 1985. When a contract expired, all outstanding positions were settled in cash at the average of the preceding five days' indices. Again, this had a somewhat cautious welcome from the trade and the market gradually faded away again.

But the pressure from the industry continued and a third attempt was made by the IPE in 1988 with the introduction of a Brent contract settled in cash but using the index for one day only rather than the five-day average used previously.

This contract was successful from the start, receiving great support from the industry and the brokerage community. Volume and open interest grew steadily, with the contract averaging more than 7,500 contracts a day by April 1989. Its volume overtook gas oil for the first time that month and it has traded more than 15,000 contracts in a day.

The concept of cash settlement was readily accepted by the oil industry and this concept was used for the introduction of a new 3.5 per cent sulphur fuel-oil contract in September 1989. There are likely to be other cash settlement contracts introduced in the near future. It is particularly useful where physical delivery provisions are impossible or awkward or where some form of hybrid contract is designed. Many physical crude-oil deals, for example, are now priced against a basket of six or seven major crudes. Such a basket could be traded on the futures market.

SIMEX fuel-oil market

In February 1989 the Singapore International Monetary Exchange (SIMEX) introduced its first oil futures contract, for high sulphur fuel oil. This had a good start. Although the volumes in the early days were boosted by high levels of cross trades, the underlying volume in the first two months was estimated at around 1,000 contracts per day, a very acceptable level for a new contract.

The contract attracted a high degree of interest before it opened because it was the first oil contract to be opened in the Far East. This led to the market being closely followed by crude-oil traders and others not involved in fuel oil itself.

SIMEX already has plans for other contracts and there are other exchanges thinking of opening oil contracts elsewhere in the Pacific region. With a physical market already trading twenty-four hours a day, the opening of SIMEX marked a move by the futures markets to reflect that. Depending on the time of year, there are now open outcry futures markets open eighteen or nineteen hours a day.

Options

There are now traded options on all the NYMEX and IPE futures contracts except the two new fuel-oil markets. Options have only been traded in the United States since 1982, when they were introduced on the financial markets, but they have already become a very important trading tool, frequently trading higher volumes than the underlying futures contract. Options will be considered in more detail in a later chapter.

The paper refinery

The introduction of crude-oil futures opened up the way to the crack spread or paper refinery. All oil refiners operate on the margin between the cost of crude oil and the value of the products produced, but when the crude oil is purchased the exact selling price of the products cannot be determined. Using the futures markets, however, refinery margins can be traded on paper. At the point where the premium of the value of products over the cost of crude makes refining economic, the product contracts are sold and the crude bought on the assumption that refinery runs will increase and the margin decline. When this happens, the crude is sold and the products bought back. If refining is far from economic, the reverse can be done, because refinery runs will fall until the margin widens.

This spread, known as the crack spread, involves having an equal number of product contracts (split roughly 2:1 or 3:2 gasoline to heating oil) to crude contracts on NYMEX or 4:3 gas oil to crude on the IPE. It remains to be seen how the split will change with the introduction of a fuel-oil contract on the NYMEX. The split reflects the yield of a standard refinery, but of necessity omits the heavier end of the barrel. Because of this omission, some traders prefer to trade a disproportionate number of crude contracts to products, for example ten crude and a total of only eight product contracts, seeing this as a more realistic reflection of the true yield. The main disadvantage of this is that the exchange's low deposits charged on spreads are not applicable unless there are equal contracts on each side.

Possible new contracts

The NYMEX plans to introduce a natural-gas contract in early 1990 although it is severely constrained by space considerations, particularly since the introduction of fuel oil trading. This will be based on the gas pipeline systems in the United States, but should attract a degree of overseas interest as well, as it will be the only contract of its type. The contract has already received the approval of the Commodities Futures

Trading Commission, the US regulatory body which has to approve all new futures contracts. In addition it may introduce contracts for heavy crude oil, fuel oil and, perhaps, Brent crude.

The IPE meanwhile is considering other crude-oil contracts, a possible gasoline contract and may enter discussions with the Baltic Exchange on a possible tanker index contract.

SIMEX would also like to introduce some new oil contracts, but the fuel-oil contract will need to be a little more firmly established before any new vehicles can be introduced. A likely contender for a new market would be Dubai crude, which is the most widely traded heavy crude and has a paper market similar to, though much smaller than, Brent.

Two new futures contracts, in gas oil and Brent, were introduced by the Rotterdam Options and Futures Exchange in October 1989. The gas-oil contract is very similar to the IPE's but the Brent contract calls for physical delivery. Early signs are that the exchange is not able to compete with the IPE on either contract. Traders seem to think that the exchange has no advantages to compensate for the inevitable lack of liquidity in new contracts.

Elsewhere, a number of exchanges are looking at the possibility of energy contracts, largely because they are the only commodity sector outside the financial instruments seeing rapid growth. There may, at some time in the future, be arrangements made for one exchange to trade another exchange's contract, in another time zone. Contracts traded could be offset against those traded on the other exchange(s) trading the same contract.

From the other side of the fence, the industry has now grown so accustomed to futures contracts that it is lobbying the exchanges for the contracts it would like to have. A heavy crude-oil contract and fuel oil probably have the strongest supporters, but there are also demands for futures markets in jet fuel, naphtha and other small-volume products. It is unlikely that a futures market in either jet fuel or naphtha would succeed in Europe or the United States, in the first case because of the unusual nature of the market and in the second because of the small number of participants. Naphtha might possibly succeed in the Far East, in either Singapore or Tokyo, because of the active market based around Japanese import requirements. It remains to be seen how well the new contracts succeed.

The demands for heavy fuel oil and heavy crude oil both stem from the increasing amount of cracking and the relative recovery in demand for heavy fuel oil. One problem encountered in establishing a fuel-oil contract is that the fuel-oil market is divided roughly into two parts: utilities and ship's bunker fuel. The former includes demand from power stations and industry. There are some significant specification differences between the two and the IPE's failure with its first fuel-oil contract was largely due to its attempt to compromise, introducing elements from both specifications to produce a hybrid contract for a product that did not exist.

There is a strong argument to be made that most oil products and some crude contracts would benefit from having a trading vehicle in each of the major time zones to provide active open-outcry trading virtually twenty-four hours a day. Ideally such contracts should be offset against each other, but rivalry between exchanges means that normally each one wants to introduce its own.

This may change with the introduction of automated twenty-four-hour trading via screens. In early 1989 the NYMEX membership agreed to proceed with plans for the introduction of screen trading, but contractual problems have prevented any real advances being made so far. When some sort of screen trading system is introduced, it will be possible to trade NYMEX futures contracts 'out of hours' by screen. Bids and offers would be displayed and the computer system would match buyers and sellers on a first come, first served basis to create an automated system as close as possible to open outcry. Such a market has already been traded. The London Fox white sugar contract, for example, is screen traded, though as an alternative to open outcry rather than as an addition to it.

The IPE is considering introducing automated trading for the period between the end of the day in London and the close of NYMEX. This would enable traders to continue to trade the London markets rather than have to take cover for IPE positions on NYMEX if there is any price movement and then move the position back to London the following day. It would also allow arbitrage to continue for two or three hours more.

Automated trading systems can also be used to trade smaller contracts, such as the London Fox white sugar contract, where the volume of business is insufficient to sustain the infrastructure needed for an open-outcry market. It might be possible, for example, to have a naphtha market operated in this way.

In addition to the design of the trading system itself, there are a number of regulatory aspects to be considered. For example, the brokers are the principals on the market, and must therefore have control of the trading done in their name, but whether this means they will be the only people allowed to have screens remains to be seen.

There seems little doubt, however, that some kind of screen trading system will be introduced over the next few years. If it is as effective as the existing open-outcry market, and more particularly if it is cheaper, it will attract a considerable amount of business. If not, after a trial period participants will return to the open-outcry market. Some combination of the two systems seems likely to provide the way forward for the futures markets.

5

Entering the futures market

The decision to trade

There are almost as many ways for the oil industry to use the futures market as there are users. Just as no two physical trading companies operate in an identical way, so too will their futures market usage vary. But most will find some occasions on which futures trading is an invaluable tool to help in limiting the risks involved in the physical market.

The decision to use futures should only be taken after careful consideration of the various methods of trading and of the operation of the markets themselves. Although the physical and futures markets move closely together, they are, on occasion, subject to different influences and prices will diverge in the short term, though they must realign in any market with a realistic delivery process. Figure 5.1 illustrates this on the gas-oil futures and gas-oil cargoes market.

Physical business can be enhanced by futures trading in a number of ways. The straight hedge allows a trader to get protection for an unattractive physical position, or a distributor can buy futures ahead and then offer a fixed price to his customers based on that price. A number of Swiss oil distributors did this in the early part of 1983 when futures prices, and physical ones, were very low. By hedging all their forward sales on the futures market they were able to lock into an effective physical price and therefore ensure a profit. Their success was such that they have repeated it almost constantly since.

To date, consumers have been less interested in the futures markets than other sectors of the oil industry, probably because price trends were generally downwards during the first few years of futures markets.

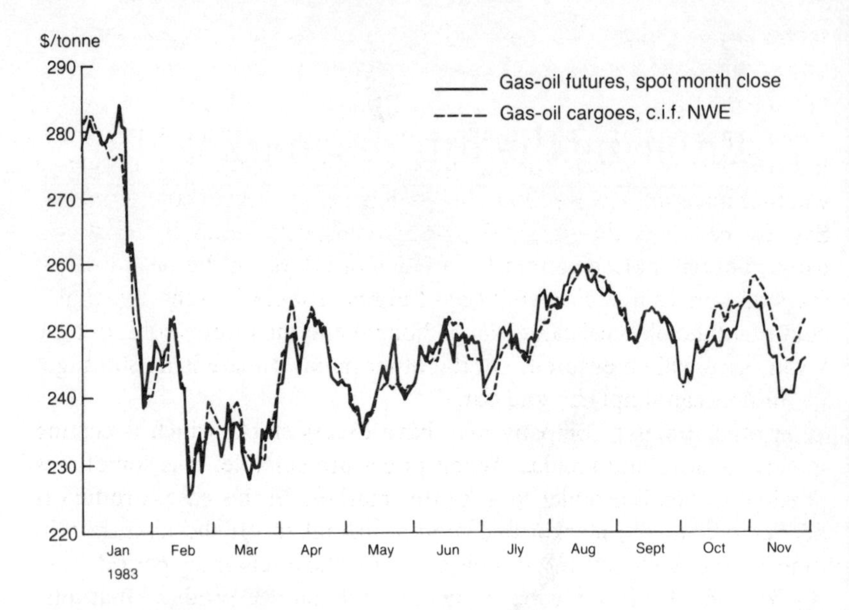

Figure 5.1 Gas-oil prices, 1983.

One major transport authority in the United States, the Washington State Transit Authority, however, bought all its oil requirements forward on NYMEX in the early 1980s, and was therefore able to guarantee ticket prices. A UK ferry company did the same thing on the IPE.

In New York and London the use of the e.f.p. (exchange for physicals) whereby a futures position is exchanged for a physical one, is widespread and accounts for most of the deliveries in any month. It is now possible to do this on all the oil futures markets, both crude and products. Details of the e.f.p. are given on page 67.

Executable orders (allowable in London but not in New York: see page 70) provide another means of pricing a physical deal on the futures market, with one side being aware of the other's price. Any difference between the two is made up by the futures profit or loss. Their attraction is that they enable both sides to choose any time, between the deal being agreed and delivery taking place, to price the contract. The danger of being caught by a temporary movement in prices is therefore considerably lower than in a normal Platts-related or similar deal based on the price on one particular day, or a range of days. A large number of spot-market deals are based on the prices reported by one of the telex price services, particularly McGraw Hill's Platts. A deal is struck with the price defined as a relationship to the published price on, for example, bill of lading day.

The liquidity of the futures market makes it very much easier to trade futures than physicals if, for example, a supply department is uncertain whether its refinery will produce enough product to meet commitments. Say, for example, the supply department suspects it may be short of a cargo, but will not be certain for a few more days. In the meantime, it fears that prices may rise, so it buys futures. Then, when the shortfall is confirmed the physical cargo can be bought and the futures sold, but the lower price will be ensured. On the other hand, if there is no shortage, the futures can simply be sold out.

At other times a company may have excess stock, which is costing money to store and finance. When prices are suitable, it is sometimes possible to 'lend' product to a futures market. In this case, product is delivered onto the market one month and taken off the next month, thus freeing up capital and storage space for the intervening period.

With the advent of a wider range of crude oil and products markets, there are also possibilities of trading the refinery margin, the so-called 'paper refinery' (see page 45), particularly in New York where the gasoline and heating-oil contracts allow for a fairly full coverage of the barrel.

The tendency of the futures market occasionally to react short term to events and influences of little or no interest to the oil industry is no reason not to use the market, but it should be understood that the two markets, futures and physical, are different, just as the barge and cargo markets are different. Any divergence should, in fact, be treated as an additional trading opportunity. The futures markets are often criticised by the oil industry for failing to react, or overreacting, to certain influences, but, although there is a speculative element in the market, the vast majority of trading is done by the oil industry itself. Studies on the NYMEX suggest that the oil industry accounts on average for about 65 per cent of trading, with the speculative element proportionately lower on the busiest days. In London the trade element is probably over 70 per cent. So although the locals, commodity funds and other speculators may distort the market for a while, they cannot control it indefinitely.

Similarly, technical trading, often dismissed by physical oil traders, can be a strong short-term influence on the markets and should not therefore be ignored. Neither, however, should it be taken as a reason not to trade the markets. All markets with a sound delivery process, be it cash settlement or a conventional physical delivery, must stay in line with its cash product in the long term. So, rather than ignore short-term influences, the physical oil trader should seek to take advantage of them.

Choosing a broker

Once a trader has decided to enter the market, the next step is to choose a broker. Each market floor has a finite number of members, and all business must be conducted through these members. In London most of the futures brokers active in the oil market have floor membership; but in New York there are a number of brokers who operate entirely from the floor, and others, with no floor operations, who have to put their business through a floor member, as well as those with more orthodox floor-and-office operations.

The choice of broker to use is a personal one, often coming down to the relationship between the individuals in the companies concerned, but there are some guidelines to use. For example, a large company trading high volumes will probably want to choose two or three brokers used to handling large orders. A smaller user may prefer a smaller broker, who will have more time to discuss the market but may charge higher commissions. Different brokers specialise in different areas of the business: there are some on both the IPE and NYMEX who specialise in serving the oil industry, and others who offer a broader investment programme for the individual speculator. In New York many of the floor members are 'locals' who do not trade client business but move in and out of the market several times a day on their own account. They can be very influential short term on the price movements, but help to provide liquidity in the market.

A company choosing two or more brokers may well decide to choose one specialising in the industry and another with good technical information in order to provide a wider view. Having too many brokers is likely to lead to confusion, but using more than one not only allows for a different opinion but prevents any one broker, however trustworthy, having a complete picture of a trader's position – a factor considered important by many companies.

Another factor to be considered is the way the futures broker reimburses its employees. Some pay their dealers a basic salary plus a profit-related bonus, while others pay the dealer a proportion of all commissions generated. There are arguments to be made for each case, but it is useful to know which system is operating for the brokers being used.

The rules and regulations covering futures trading (apart from the market rules) are somewhat different in the United States, the United Kingdom and elsewhere. In the United States, the Commodities Futures Trading Commission (CFTC) is the government agency concerned with futures trading and all brokerage activities are subject to its rules. All brokers are also members of the National Futures Association (NFA), the self-governing regulatory body which deals with futures broking. The NFA is financed by a levy paid on each futures contract traded on a US exchange.

In the United Kingdom, the Securities and Investments Board (SIB) is the umbrella organisation dealing with all forms of investment activity and the Association of Futures Brokers and Dealers (AFBD) is the self-regulatory organisation governing futures broking. They are both financed by charges made on their members.

Elsewhere, regulation is being developed, usually along the lines of the US or UK legislation. In general, regulations imposed in any country govern all business carried out on exchanges in that country or by brokers working in that country. Although regulation differs from place to place, the general principles are more or less the same: to ensure good financial and trading practice within futures brokers and consequently a high level of service to customers.

One of the main differences between the US and UK legislation is that US brokers are not allowed to give credit to their customers, while UK brokers are. There are controls and monitors imposed by the SIB and the AFBD on the amount of credit that can be granted: these are designed to ensure that no UK broker becomes so exposed to one customer that that customer's failure to perform can force the broker out of business or prevent his fulfilling his obligations to other clients.

Another is that US brokers are obliged to keep all client monies in separate bank accounts from the broker's own funds. In the United Kingdom, the client is able to choose whether he wants his funds segregated from the broker's or not.

Clearing

Every futures exchange has a clearing mechanism which guarantees all the trades on the market, once they have been correctly registered. The clearing house registers all trades made on the exchange floor, allocates them to the members and effectively steps in as buyer to every seller and seller to every buyer. A trade is executed on the floor between two brokers but as soon as it is correctly registered with the clearing house it becomes two separate positions held with the clearing house. In this way, when a closing trade is made, the position with the clearing house can simply be closed without reference to any other broker's position.

All exchanges have clearing members, who are authorised to hold positions on that market with the relevant clearing house. These members are not necessarily futures brokers, but may be large users of the market. They are not necessarily present on the floor of the market, and neither are all floor brokers necessarily clearing members.

All trades on a market must be executed through a floor broker. These are the brokers actually on the floor of the market authorised by the exchange to trade. A large proportion of floor members are also clearing members and so can hold their clients', and if applicable their own, positions. Others are not and have to 'give up' all trades to a clearing member. They are paid a fee for their floor execution.

Clearing members are responsible to the clearing house for the trades registered in their name. Unlike some other types of market, futures brokers are the principal to the market. This enables clients to maintain anonymity, but means that care should be taken to check the financial viability of the futures broker before opening a trading relationship. This is one of the reasons why some large users of the market prefer to become clearing members themselves. Although they then have to pay membership fees and subscriptions, and on some markets buy seats on the floor, they usually enjoy lower clearing fees and also have the facility to trade with a number of brokers but only have one overall position.

The clearing houses differ from market to market. The IPE is cleared by the International Commodities Clearing House (ICCH), an organisation owned by the major UK clearing banks. Both SIMEX and NYMEX are cleared by what are effectively mutual corporations of their members, supported by the financial resources of those members.

In order to provide the financial guarantee, the clearing house charges an initial margin or deposit on each lot held overnight. These initial margins are set by the exchange/clearing house and can be varied at any time if the clearing house believes market conditions warrant either an increase or decrease. In mid 1989, initial margins were $2,000 per lot on NYMEX, $1,000 per lot on the IPE and $500 per lot on SIMEX.

Each day, all futures positions are margined in full by the clearing house. The full difference in value between the price at which the futures contract was bought or sold and the market price at the previous day's close is payable to the clearing house if negative and to the client if positive. The amount of this difference is called the variation margin.

Commissions are payable to futures brokers for business transacted. The level of the commission has to be negotiated and will depend on the volume of business being done and the service required from the broker. For example, some clients like to choose their own floor brokers while others execute and clear through the same broker.

Futures commissions differ from physical market commissions in that the futures broker has to pay market fees and clearing fees for each lot traded and does not therefore retain the full amount. It should also be remembered that the broker is the principal to the market and is therefore assuming market risk whenever it trades for a client.

In the United States, futures brokers are required to submit a report to the CFTC each day showing all positions of more than twenty-five lots in any one month held by their clients. There are various restrictions on the total position any one client can hold on the market. Some of these restrictions can be eased upon application from the client to the CFTC, which requires evidence that a company needs to hold large futures positions to hedge its physical positions.

An oil company trading on the markets will probably find it helpful to discuss with its broker what its purpose is in trading in futures. It is not necessary to give too many details, but a general outline will enable the broker to give better advice. It must be remembered that, much more than the physical market, there is great secrecy surrounding futures activity. No broker divulges the names or trading position of a client, so the client will not suffer from telling the broker whether a proposed trade is a hedge, a speculative trade or whatever. Although a broker will not be able or even wish to change a client's mind about the position, he will be able to give advice on timing; the state of the market at any given time (for example, it can be almost impossible to trade large volumes during quiet periods); and likely short-term developments.

A technical trader, who bases his trading system on various charting methods, will not look for any advice from his broker – indeed a true chartist should take no notice whatsoever of any fundamental information which may be offered.

In both London and New York the oil futures markets are used to an unusual degree by the industry itself, as opposed to general investors and speculators. Recent studies in New York suggest that the oil industry accounts for 60–70 per cent of the activity in heating oil and gasoline, and around 70 per cent in crude oil, while London estimates trade participation at 65–75 per cent. It is likely that the crude-oil market will eventually attract more non-trade business than the products market because of the intrinsic appeal of crude oil, like gold, to the speculator.

Despite this high level of trade involvement, the markets have technically behaved 'well', encouraging the commodity funds to come into the market. The oil trade sometimes worries about the activity of speculative traders in 'its' market, but it must be remembered that the speculators help to take on the risk the oil industry is trying to lay off, and improve liquidity.

One major attraction to chartists is that the oil price over the last six or seven years has tended to fall into fairly long-term (six to ten weeks) trends. There have been few occasions when the market has seen temporary changes of direction apart from technical reactions to sudden moves, which in any case tend to get ironed out in chart assessment. There have been several occasions, however, when the market has barely moved for several weeks, which tends to discourage speculators who require a shifting market.

In-house administration

Almost as important as the choice of broker is the setting up of effective in-house administration systems. Several European and US oil traders have suffered from having their futures operations physically separate from their spot-market activities. Although it may be necessary for different individuals to take responsibility, it is not practicable to run a futures book entirely separately from a physical book as effective use of the futures market necessitates close interaction between the two.

The mechanics of futures trading

Although exchange practices and regulations vary from market to market, there is one thing that all futures markets have in common. This is the conduct of trading, except of e.f.p.s, by open outcry between a restricted number of members on a trading floor between certain hours. In London all floor members are companies, required by IPE regulations to meet certain capital and other financial criteria, but in the United States memberships are held by individuals and used either by their companies or by the individuals themselves. These individuals are known as 'locals' and trade primarily on their own behalf, moving in and out of the market several times a day and helping to provide liquidity. They also execute client orders.

The open-outcry system reflects the origins of futures markets in corn exchanges and similar open markets, but is thought by some to be inadequate for the high-volume markets seen today, particularly in the financial instruments. Hence the development of automated screen trading systems.

Under the open-outcry system, bids and offers are shouted across the floor of the market until agreement is reached. The two dealers concerned then agree the number of lots traded and the deal is registered and reported back to the client.

Mechanically, futures trading is simple to operate. There are several ways of giving an order to a broker, but in all cases the order is executed on the floor of the market and then registered and cleared in the normal way. The means of confirming the order to the client vary a little but are normally determined by the client. Market practice is normally a telex confirmation of all executed orders (except in the United States) and written confirmation the following day. Each time a position is closed out, a settlement contract is sent out and then each month a summary of trades and an open position statement are sent.

There are various ways of entering an order to a futures broker, depending on the result required. The most common order is one to buy or sell a certain number of lots at a certain price. A slight variation, particularly for a larger order, is to buy or sell up to or down to a certain price slightly above or slightly below the prevailing market price.

There are also a number of other types of order, used in different circumstances. These include an 'at best' order, where the client asks the broker to use his discretion to obtain the best price. The client is not guaranteed a fill using this type of order. If the broker, using his discretion, decides not to trade, the client has no right to a fill. Normally, if a client has put in an order at a price and the market trades below his price on a buy order or above it on a sell order, he is guaranteed an execution.

An order can also be given 'at market'. This requires the broker to go into the market and buy or sell at the best price he can at the time. This type of order is most often given in a fast moving market where a client wants to get into the market and is more concerned with getting his position on (or closing it out) than the last point or two on the execution. It is in every broker's interest to try and get the best fill he can for his client on any discretionary or market order, and once a relationship of trust has been established the client will often find it advantageous to give some discretion to the broker.

'Stop' or 'market if touched' orders are used to limit losses or to enter the market on a technical basis. These are orders placed at a certain level but which become market orders once that level is reached. Thus if the market trades at a particular price the order will be executed immediately, but not necessarily at the price given.

There are also various other orders such as 'market on close', a market order executed only during the close of the market; 'o.c.o.' (one cancels other) where two orders are entered together, usually a stop order and an ordinary one, and as soon as one is executed, the other is cancelled; 'or better' or 'not held' where a level is given but the broker is allowed to hold back if he thinks he can do better, but no fill is guaranteed; and 'good till cancelled' (g.t.c.) where the order is left in the market until it is filled, however long that may be – normally orders are cancelled at the end of the day, but g.t.c. orders remain indefinitely until filled or cancelled.

6

Strategies in futures trading

In this chapter we will look at the different ways the futures markets can be used by the various sectors of the oil industry. The oil futures markets have, in their relatively short history, exhibited a close relationship with the physical markets they serve. Inevitably, the two move apart from time to time, but this should not deter the oil man from using the market. Rather it should encourage him to look for further trading opportunities.

This chapter deals only with futures markets. Options are discussed in Chapter 7.

Hedging

The most talked-about form of futures trading is the hedge – the taking of a futures position equal and opposite to the physical position to be protected. A perfect hedge is impossible to achieve, because of the minor variations between the futures and physical markets, but the relationship is certainly close enough to make hedging attractive.

Probably the most common hedge is one taken against an existing physical position actually held. For example, a trader is long of a cargo of gas oil but is somewhat nervous of the price trends on the physical market. But he is unable to sell his cargo immediately, so he sells futures instead. Then, when the physical cargo is sold, the futures are bought back.

The hedge can also be used to fix prices for future transactions on the physical market. A gas-oil consumer, for example, knows that he will have to purchase gas oil in October, but anticipates rising prices. So, earlier in the year, he can buy futures. Then, when he takes delivery of his physical product, he sells his futures. Or a refiner, knowing he will be producing a certain amount of product, can, having established his crude price, sell forward to guarantee a profit.

Hedging can be split into two categories, the short hedge (involving the sale of futures) and the long hedge (buying futures).

The short hedge

The most frequent users of the short hedge are: a trader long of oil but anticipating a fall in the price; a refiner who knows he will be producing a product but expects prices to fall; a refiner who wishes to lock-in a profit on processing his crude; a crude-oil producer; a supply department with possible excess product to sell, but which is unwilling to sell on the physical market, either because of price or because the product may not be forthcoming.

Futures markets are based on the idea of a standard product in standard quantities. It may not be possible to match absolutely either size or type of crude oil or product, but the nearest possible match should be aimed for. The following example illustrates this.

Example

A trader has bought a cargo of 25,000 tonnes of gas oil, but has now decided that the market is less steady than he had believed. He has already agreed the sale of the product, on a Platts-related basis, for ten days ahead. He therefore sells futures contracts to protect against the anticipated fall in prices. When the physical cargo is sold, the futures are bought back.

	Physical		Futures	
		$/tonne		$/tonne
2 July	Long 25,000 tonnes at	158.00	Sells 250 lots at	163.00
13 July	Cargo sold at	142.50	Buys 250 lots at	149.00
	Physical loss	15.50	Futures profit	14.00
	Net trading loss is $1.50/tonne			

Thus the trader has protected himself against the fall in price seen between buying his cargo and selling it. Although the hedge was not perfect, it saved him $350,000 or the problems involved in selling the first cargo and buying again to meet his commitments. The cost of the futures transaction is variable, depending on the terms agreed with his broker, but would almost certainly be less than $0.30/tonne, including the costs of margin payments, deposits and commission.

Some of the other possible users, such as the supply department uncertain whether or not it will have excess product, are attracted to the hedge because it is easy to get out of. So, in the case of a supply department with possible excess product, if the refiner calls to say that production will be lower than anticipated, it is quickly possible to remove the hedge, or even turn it round if the position is reversed.

Once a hedge is put on, it does not necessarily have to be left untouched until the physical position is changed. It is possible to trade inside the hedge, perhaps buying some of the contracts back earlier and then selling them again if the price goes up. In this case, there is some speculation involved, as the futures position is no longer equal and opposite to the physical position but is being changed to reflect a view of the market.

The long hedge

This can be used by a trader short of product; a distributor with future commitments; a consumer; a supply department short of product; a refiner wishing to lock into a crude-oil price, etc.

Example
A consumer knows that he will take delivery of 10,000 tonnes of product in two months' time, but thinks that the price will rise in the intervening period. He is unable to buy the product now, because there is no room to store it yet. So, he buys 100 lots of gas-oil futures on the IPE, selling them out when his physical delivery is made.

	Physical	Futures
August		Buys 100 lots October at $164.00/tonne
October	Buys 10,000 tonnes at $194.00	Sells 100 lots October at $192.00/tonne
		Futures profit $28.00/tonne

Net cost of physical oil is $166.00/tonne

> Again, the cost of the transaction will be relatively small though the fact that the hedge is held for two months makes the interest paid or assumed, on the deposit and the margin payments, greater.

The costs of dealing on the futures markets are explained in some detail in Appendix B. In brief, however, they depend on the terms agreed between client and broker. An assumed interest, based on loss of interest on capital which would otherwise be used elsewhere, should be used in calculating the cost of trading. Certain other forms of security, such as Treasury bills, can sometimes be used in payment of margins and deposits.

Switches

Switches, also known as spreads or straddles, involve the simultaneous purchase of one contract and sale of another, in a different month or different product, to trade the differential. The actual price of the contracts becomes irrelevant, it is only the differential which is of interest.

The introduction of new futures contracts is increasing the scope of spread trading enormously. With crude oil and several products now available, it is possible to trade the crack spread, based on refinery economics, as well as the difference between the London and New York gas-oil and crude-oil markets, two months within one product or two different products.

The simplest switch is that in which the same market is bought for one month and sold for another. This is most often done as a speculative trade, when one month appears to be getting out of line with another in the view of the trader; but it can also be tied in with a company's physical business.

'Carrying' switches

From time to time most futures markets see the carrying charge reflected in the prices trading in different months. In other words, the price of a forward month is greater than the price of a nearer one plus the cost of keeping the product until the forward month. When this happens, the nearer month is bought and the further one sold. The deal is closed-out by taking delivery of the product off the market and putting it back in the further month. The cost of the carry is dependent on the cost of storing the material, the cost of financing the purchase and the cost of taking and making delivery on the market.

Effecting a cash-and-carry is virtually the same on all the oil product markets, and on crude oil although the independent storage of crude oil is less common than of oil products. In all cases, calculations of costs must be made based on the worst possible case. For example, a buyer taking delivery of gas oil, heating oil or gasoline from the futures markets must nominate a five-day delivery range for collecting the material, but the first nomination may be rejected by the seller. Ideally, doing the cash-and-carry, a buyer would nominate the last five days of the month, but rejection by the seller would mean the latest the product could be transferred would be the 25/26 day of the month. (Late transfer gives the shortest time between payment for the products and redelivery.) The same rule must be considered on redelivery; the buyer may nominate the last delivery period, which can be substituted by the seller, but again transfer may not happen until the 25/26 of the month.

On the NYMEX and IPE product markets, inter-tank transfer is a standard method of delivery, but in both London and New York it may be necessary to remove the product from the installation where delivery has taken place, adding enormously to the costs. This is likely to be more of a problem on the IPE because of Dutch laws preventing refineries storing third-party material.

Storage rates vary from installation to installation, so again the worst case must be assumed. Although the buyer on both NYMEX and the IPE may express a preference for delivery location, the choice is the seller's. A typical New York Harbor storage rate of 2 cents per gallon is used in this example.

Example

		cents/gallon
April	May gasoline bought at	76.00
	June gasoline sold at	80.00
4 May	Delivery range 22–26 May agreed	
26 May	Delivery made by inter-tank transfer	
27 May	Payment made	76.00
	Storage for one month paid	2.00
		78.00

	Interest for one month at 10.00% per annum	0.65
	Total outlay	78.65
25 June	Delivery made	
26 June	Payment received	80.00
	Gross profit	1.35
	Commission	0.05
	Net profit	1.30

The carry can also be worked in reverse, when product is lent to the market for a month (or more) because the difference in value more than reflects the cost of making delivery. This is a less popular form of trading, because it is dependent on the ability actually to deliver oil, but can be very useful for companies with oil stocks.

In executing either of these carrying spreads, care must be taken in making calculations and, if any volume is being traded several months forward, the interest rate should be watched closely. It is not often possible to make large profits trading in this way, and a change in the interest rate (or the exchange rate between the Dutch guilder and the dollar) can soon wipe out a modest return.

Crude–product spreads

With the advent of more futures contracts, the possibilities for switch trading are opening up. In New York there is a separate ring for trading spreads between the different contracts, and the exchange quotes settlement prices daily in the same way as for normal contracts. Not all spread trading is done in the spread pit, much is still traded in the individual contract rings. When a client gives a spread order, it is up to the broker which way it is executed – it can be riskier for him to use both markets simultaneously because if he is only able to do one side and the market moves before he has done the other, he obviously cannot give the trade to the client. But he may be able to get a slightly better price for his client by 'lifting legs'.

Crack spreads are a very popular trading vehicle. The most common ones are those involving one product, either heating oil, gas oil or gasoline, against crude, but there are also a large number of 3:2:1

(crude : gasoline : heating oil) and 5:3:2 spreads traded. In both cases, exchange deposits are reduced.

It is impossible to get a perfect match to refinery output, the basic theory behind crack spreads, because of the lack of active fuel-oil contracts and the absence of a gasoline contract in London, but the general principles remain the same. Some traders will trade crude contracts for one month against product contracts for the following, to take account of the time lag between a refiner's buying crude and having product to sell, but most tend to trade the same months.

The trading of two products against crude is usually a more reliable indicator of refiners' intentions, because it gives a truer reflection of the overall state of refinery economics. The boundaries of the spread's movement are harder to determine than, say, those of a cash-and-carry spread. For example, during the summer of 1988, refinery utilisation was very high and the crack spreads between gasoline and crude continued to widen beyond what would normally be considered a high margin. But the refiners were not able to increase output and the spreads continued to widen.

When the premium of products over crude becomes too wide, the futures trader would therefore sell the products and buy the crude. It does not matter to him whether the differential narrows because crude prices rise or product prices fall, provided that it *does* narrow. As with other spread trading it is the differential which is of importance not the actual price.

When the crude becomes overvalued with respect to the products, the reverse will be done – the crude sold and the products bought.

Example

NYMEX gasoline is trading at 71.50 cents/gallon, heating oil at 51.00 cents/gallon and WTI at $20.25 per barrel, giving a differential of $6.91 per barrel. The products are sold and the crude oil bought, then, when the differential narrows, the position is lifted.

Three crude oil bought	$20.25 per barrel
Two gasoline sold	71.50 cents/gallon
One heating oil sold	51.00 cents/gallon
(differential $6.91 per barrel)	

Later:

Three crude oil sold	\$20.00 per barrel
Two gasoline bought	68.00 cents/gallon
One heating oil bought	52.00 cents/gallon
(differential \$6.32 per barrel)	

The profit on the trade would be \$0.59 per barrel less six commissions totalling \$0.12 per barrel at \$20.00 round turn per lot. Margins would also have to be paid.

As with all other spreads, the actual price levels of the different contracts do not matter, it is the differentials between them that are important. In the above examples the same result would have been obtained if the position had been lifted when crude was trading at \$16.50 per barrel, gasoline at 60 cents/gallon and heating oil at 43 cents/gallon.

Arbitrage

Arbitrage is the trading of the price differential between two markets for the same or similar product: it can mean the difference between the cash market and related futures market (particularly in the United States) but here we will only deal with the difference between the same product in different futures markets. With the introduction of an increasing number of oil contracts around the world, the possibilities for arbitrage trading are increasing.

The most common arbitrages are those between NYMEX heating oil and IPE gas oil and between WTI and Brent. In order to trade the arbitrage between any two markets it is necessary for them both to be open at the same time: it would not therefore be possible to arbitrage between Singapore fuel oil and the fuel-oil contract introduced in New York: there could, however, be spread trading between the IPE and Singapore, because there is short overlap.

Care must be taken when arbitraging between two markets to ensure that the volumes traded in each market are the same. For example, when trading the heating-oil/gas-oil arbitrage, four gas-oil contracts are traded for every three heating-oil contracts. This arbitrage also involves comparing the price of one contract priced on a volumetric basis and one on a weight basis. Different traders use different conversion factors, but perhaps the most common is that based on the IPE gas-oil standard density.

Example
NYMEX heating oil is trading at a premium to the IPE gas oil of $5.00 per tonne. A trader anticipating a widening of this differential would buy heating oil and sell gas oil, being careful to maintain the correct number of contracts (three heating oil to every four gas oil). Different traders use slightly different conversion factors to obtain a heating-oil price in dollars/tonne (for Europeans) or a gas-oil price in cents/gallon (for US traders). Probably the most common conversion factor is 313 gallons/tonne, the number at the IPE gas oil's standard density of 0.845 kgs/litre. By multiplying the heating-oil price by 3.13 the equivalent dollars/tonne price is obtained, and by dividing the gas oil price by 3.13 a cents/gallon price is reached.

	$/tonne	cents/gallon
Four gas oil sold	154.00	49.20
Three heating oil bought	159.00	50.80
Differential	5.00	1.60
Later:		
Four gas oil bought	148.00	47.28
Three heating oil sold	158.00	50.48
Differential	10.00	3.20

The profit is therefore $5.00 per tonne or 1.60 cents/gallon, less seven commissions totalling $1.40 per tonne or 0.45 cents/gallon. The net profit is therefore $3.60 per tonne or 1.15 cents/gallon.

Futures strategies for physical trading

We have dealt so far with methods which can be used for both speculative and physical business. There are two types of trading, executable orders and exchange for physicals (e.f.p.), which are specifically tied in to physical business.

Exchange for physicals

An e.f.p. is the exchange of a futures position for a physical position. In effect, the two parties convert their futures hedges into physical positions. Sometimes it is simply used as an alternative method of delivery, but more commonly the physical transaction is agreed before the futures market position is taken.

The major attraction of an e.f.p. is that it separates the pricing of the oil from the physical supply. All the details relating to physical supply are agreed between the two parties, but it is agreed to price against an agreed futures market contract. The month to be used on the market must be determined, as must the date on which the e.f.p. will be registered and the number of lots to be traded.

The e.f.p. will normally be registered on or near the day the physical transfer takes place. It must be registered before the contract expires on the IPE or up to 12 noon the following day on NYMEX. Registration simply means the execution on the floor of the transfer of positions and the price at which the transfer is made.

The volume to be traded will normally be the agreed physical volume. Tolerance arrangements must be made separately. Some traders will e.f.p. the total volume and then price the tolerance at the e.f.p. price (plus or minus differential). Sometimes the e.f.p. will be registered only after the actual delivery volume is known and the exact amount will be exchanged with buyer or seller taking the tolerance risk as normal (and hedging it if thought desirable).

It is also important to establish a differential or a means of agreeing a differential if the product or crude being delivered does not exactly match the futures contract. Sometimes the differential is agreed at the same time as the physical transaction, if it is relatively constant; more normally it is agreed close to the delivery, possibly using an average of a published differential over a few days. These differentials rarely cause too many problems in practice because most differentials are fairly transparent at any one moment although they vary over time. Failure to agree a differential will result in one party having a long futures position, the other a short one and no physical deal taking place.

Once the deal has been agreed, each party is free to use the futures market at any time he wishes, trading in and out of his positions whenever he feels appropriate. On the agreed day both parties notify their brokers that the e.f.p. is to be registered, giving the number of lots and the price at which the e.f.p. is to be made. This price is used to close out the futures position and is also the invoice price (plus or minus differential) for the physical deal. It is not, therefore, significant. Most e.f.p.s are registered at the previous night's close or the current market price.

The registration of an e.f.p. does not have to close a trader's futures position – it can create one. A physical seller, for example, who will receive long contracts with the registration of an e.f.p., may prefer to take the long position and sell out when prices rise.

The registration of an e.f.p. is the only time the futures markets can be traded without open outcry. Only the number of lots and the month traded are disclosed to the market, though the exchange and the clearing house have to be told the price. The price does not have to be within the day's trading range and is not important to the exchange. What is important to the exchange, however, is that e.f.p.s are only done as part of a physical transaction. The exchanges are able to, and do, ask for documentary proof of the physical transfer.

It is important to note that it is only the futures side of an e.f.p. which is guaranteed by the exchange and clearing house. An e.f.p. is essentially a physical transaction and should therefore only be entered into with those other parties a trader is authorised to deal with.

Example

Company A is selling 450,000 barrels Ninian crude to Company B using the August Brent contract to price. The e.f.p. is to be registered on the last trading day, July 10, at the previous day's settlement price and the differential is to be the market rate on July 10.

Company A builds up a short position at an average price of $16.75 per barrel and Company B a long position at an average price of $15.60 per barrel. The settlement price on July 9 is $16.00 per barrel and the differential on the day is 10 cents discount.

Company A		Company B	
	$/barrel		$/barrel
Short futures	16.75	Long futures	15.60
E.f.p. registered	16.00	E.f.p. registered	16.00
Futures profit	0.75	Futures profit	0.40
Invoices B	15.90	Invoiced by A	15.90
Futures profit	0.75	Futures profit	0.40
Net sales price	16.65	Net purchase price	15.50

Thus it can be seen that both buyer and seller have a net price of their futures price less the discount. Whatever the registration price is, the result will be the same.

All these methods of using futures markets are being constantly refined and adjusted by the industry to adapt to their needs. Most of the new trading methods seen on the oil market in the last few years have resulted from these methods and the options strategies discussed in Chapter 7.

Executable orders

When an executable order is used, a physical deal is arranged between the two parties in the normal way. A contract for the deal is drawn up, which is exactly like a normal contract except for the pricing and payment clauses. The first of these clauses is written to say that the price will be established by executable orders, for a given number of lots, submitted to the same futures broker by a specified date. Pricing adjustments for quality can also be incorporated, as can the means of pricing any quantity adjustments within the given tolerances. Payment is usually made through the futures broker concerned.

In practice, the oil industry has slightly modified this procedure, which has become known as trigger pricing. This is more or less the same as an executable order with either buyer's options to price or seller's options to price. In the case of a seller's option to price, for example, the seller will call the buyer when he wants to price part of the contract. The buyer then accepts that price and decides himself whether he wishes to enter the futures market and take the position or to assume it himself. This method is used extensively in the market. Often there is a minimum size for each pricing decision, say 50,000 barrels on a crude contract. This means that when the buyer or seller elects to price, he must price one or more 50,000 barrel 'pieces'.

Example

Company A agrees to sell Company B 10,000 tonnes of gas oil for delivery in early November at a price fixed against the November futures contract using Broker C. All futures transactions must be completed by 31 October.

In late September Company A believes prices will begin to

weaken, and sells all 100 futures contracts at $160 per tonne. Company B has so far taken no action. A week or so later, he decides that, although the price may weaken further, he is worried by Middle East tensions, so he buys 50 lots at $152 per tonne. Two weeks later still, the price has hardly moved and he decides to buy the rest of his contracts at $151 per tonne.

On 31 October Broker C closes out A's short position against B's long position, giving a profit of $8.50 per tonne.

Thus, the balance sheet looks like this:

	$/tonne
Company A short at	160.00
Company B long at	151.50
Futures profit	8.50
Broker received from B	151.50
Adds futures profit	8.50
Broker pays A	160.00

Both sides then pay a commission to the broker, who makes no other money out of the transaction.

This type of trading can be of use to companies trading either immediately or months ahead. By using the futures market to price the deal neither buyer nor seller will suffer from any temporary aberration in the price of gas oil; indeed one side may even benefit from it.

Executable orders can be used by two traders or by a consumer in combination with a distributor, producer or other source of supply.

7

Options

Traded options are a fairly recent development in the oil futures markets, but have quickly become an integral part of many companies' trading strategies. They open up a whole new range of trading possibilities and can be used to take advantage of any view of the market – even a view that it will move sideways.

This chapter is only meant to give an outline of option theory and strategies. Those wishing to study the subject more deeply will find a wide range of specialist books available.

Firstly, it is important to note that options, despite the limited risk they offer buyers, are not always the right instrument to use. They have a great number of uses and provide opportunities and protection not available elsewhere. But to trade options profitably it is still necessary to have a view on the market and an objective.

An option gives the buyer the right to be long (a call option) or short (a put option) of a futures contract at a specified price (the strike price) in return for the payment of a premium. The buyer has until the option expiry date, or declaration date, to decide whether or not to exercise his option. The declaration date is normally a few days before the expiry of the underlying futures contract.

Having paid the premium in full at the time of purchase, the buyer does not have to take any further action, or pay any margins, unless or until he decides to exercise his option. On exercise, an option becomes a normal futures contract and is margined accordingly. If a buyer decides not to exercise his option he need do nothing and his loss is limited to the premium he paid.

The seller of the option, on the other hand, receives the premium in full at the time of sale, but has to pay initial and variation margins. His risk is unlimited, hence the margin requirement, while his profit is limited to the premium.

A buyer will only exercise his option if it is profitable or 'in-the-money'; that is if the market is above the strike price of a call option or below the strike price of a put option. When the market is at the strike price, the option is called 'at-the-money'. An 'out-of-the-money' option is one which would be unprofitable if exercised, that is a put option with a strike price below the market or a call option with a strike price above the market.

Traded options are sometimes called American options to distinguish them from the older European options which could not be traded after purchase. The length of these options is negotiable, not fixed by the exchange as with traded options, but the option cannot be sold back or bought on the market. Although these options still exist on some exchanges, they are not available on any current oil contract. Traded options are fully tradable instruments and are cleared in a similar way to the futures markets, though the margining is slightly different. Thus anyone who has sold an option can buy it back at any time and any buyer can sell it out in an open-outcry market similar to a futures market.

The option premium is made up of three distinct parts: time value, intrinsic value and volatility. (In some markets other than oil interest rates are also a factor.) Time value is determined by the time to the expiry of the option and the intrinsic value is determined by the difference between the strike price and the current market price of the underlying futures contract. An in-the-money option is inevitably more expensive than an out-of-the-money one and an option with a long time to run more expensive than a short-term one.

The biggest variable in the premium is the volatility. There are two measures of volatility – historical and implied. Historical volatility is calculated from past price movements and is the annualised standard deviation of the price of the underlying futures contract. It can be used to calculate the theoretical value of an option premium.

More important, however, is the implied volatility, which is calculated back from the option premium and is a measure of the price movement the market expects. Options are traded in an open market and are therefore subjected to the normal market laws of supply and demand, and the additional influences of sentiment. Actual option premiums therefore deviate significantly from the theoretical values.

It can happen that, in the run up to an OPEC meeting for example, the market is moving generally sideways and the historical volatility is falling but the implied volatility is increasing because the market is expecting a major price move.

Sometimes the implied volatility of put options can differ markedly from that of call options. This will happen when traders feel more need to protect themselves from a move in one direction than the other. Before the OPEC meeting at the end of 1988, for example, the implied volatility of put options was considerably higher than that of calls because traders felt, wrongly as it turned out, that a downward move was more likely than an upward one and had been buying significantly more puts than calls. Their risk was, however, limited to the option premium paid, whereas if they had sold the market outright they would have had unlimited risk as the market moved higher.

Very few options are exercised. Most are not held until expiry but are sold or bought back on the market. On the last day of trading the premium of an in-the-money option is the intrinsic value and by selling an option out the trader avoids the initial and variation requirements of a futures position, does not have to bear the risk of the position overnight and probably saves on broker's commission.

Selling options

Most option strategies are generally looked at from the buyer's point of view because it is possible to quantify his risks and to see his potential profitability at any time. As far as the seller is concerned, he can never make a greater profit than the premium he has received. If the option is out-of-the-money it will not be exercised and the seller will have made his maximum profit, the total premium. If not, the seller's potential loss increases as the option moves further into the money. An option seller will normally have physical cover for his sales or will take cover on the futures market or buy the option back if the market moves through his strike price.

Making any trade which has limited profitability and unlimited risk may seem unattractive, but there are plenty of occasions when selling an option has considerable appeal for a physical oil trader. For instance, a producer of crude oil or products has inventory and may like to sell options to generate an income from it. If prices are, in his opinion, high he may decide to sell call options with a strike price above the current market value. If he is right and the market moves lower, his option will decrease in value, giving him a profit to offset against the cost of his inventory. If he is wrong, he can buy the option back or buy futures or use his physical material as cover. He is, in effect, setting a maximum selling price for his oil.

An option seller will always gain from the decline in time value, which declines as an option approaches expiry. This decrease in value becomes greater towards the end of an option's life as each day represents a greater proportion of the remaining time.

There are a number of specialist market-makers on all the options markets, trading options on a mathematical basis. They are looking for discrepancies in the price of different options and in the different volatilities. They are active in the underlying futures contracts, constantly taking cover and building up option/futures strategies.

Option strategies

The most simple option strategies are buying puts and calls. Strategies can become as complicated as traders wish to make them involving any number of options, futures and physical legs. When working out some of these strategies, however, it is always worth checking to see that there is not a simpler way of achieving the same objective. Until a trader is totally familiar with the subject, it is often useful to break a strategy up into its different components and work out what the result would be at different price levels.

The diagrams in this section show the profit/loss profile of the different strategies at expiry with the buyer's profile shown by a solid line. The seller's profile is a mirror image of the buyer's.

A put option gives the buyer the right to be short of the market at the strike price. The trade becomes profitable as soon as the market has fallen below the strike price less the premium paid. Thus a buyer of a June WTI $18.00 put option for 45 cents begins to see a profit as soon as the market falls below $17.55 (see Figure 7.1). The buyer would, however, sell or exercise his option if the market were anywhere below $18.00 at expiry to recover some of his premium. He may of course sell it (or, unusually, exercise it) at any time between purchase and expiry. It is not usually a good idea to exercise an option before expiry because the buyer is then giving away the unexpired time value.

Similarly a call option, giving the buyer the right to be long of the

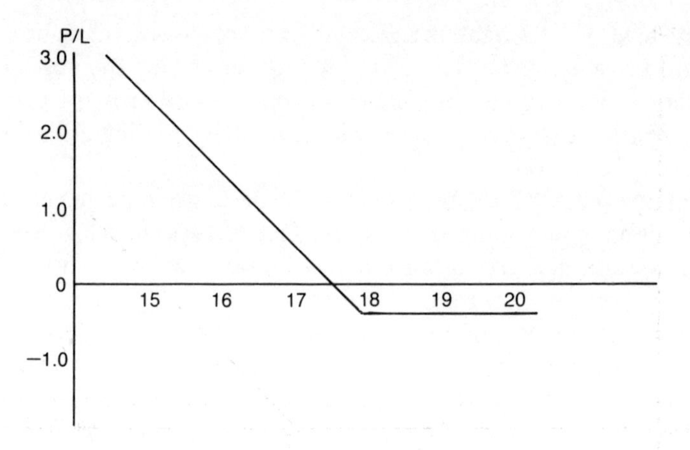

Figure 7.1 Put.

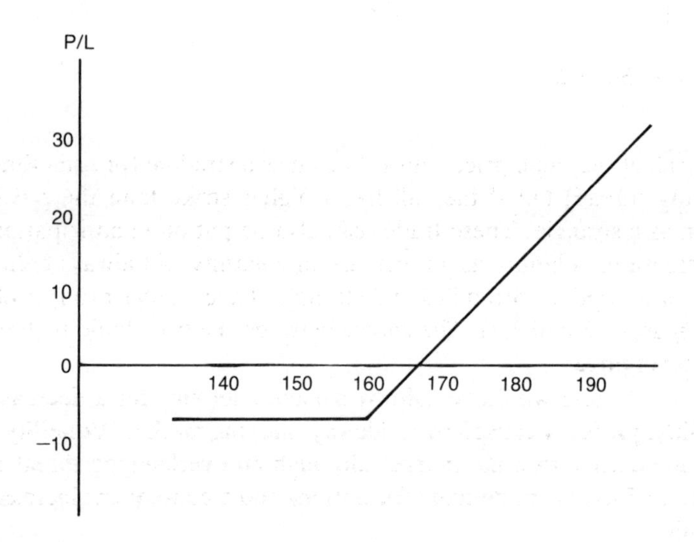

Figure 7.2 Call.

market, becomes profitable when the market moves above the strike price plus the premium. So a March $160.00 gas-oil call option bought for $7.00 becomes profitable when March futures move above $167.00 but the loss begins to reduce after $160.00 (see Figure 7.2).

A trader looking for a significant move, but uncertain as to the direction of it, may decide to buy both puts and calls. If he buys a put

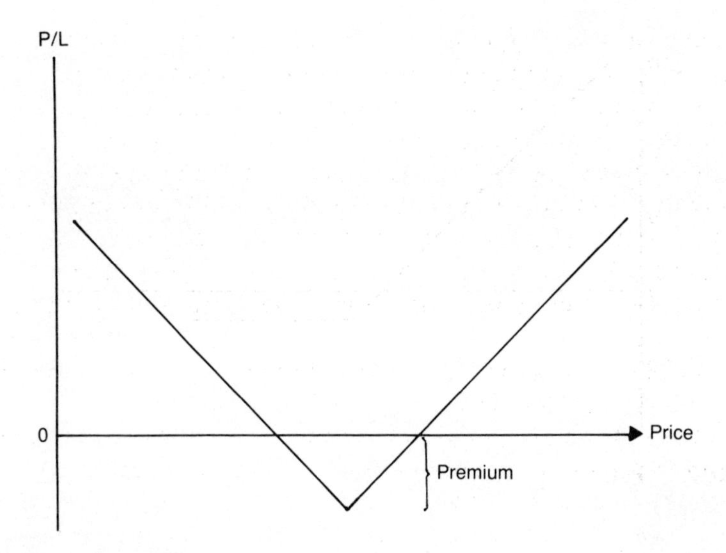

Figure 7.3 Straddle.

and a call at the same price, this is known as a straddle (or sometimes a volatility spread) but if the call has a higher strike than the put it is known as a strangle. These trades can also be put on in anticipation of an increase in volatility as an increase in volatility will always help the buyer of an option. Straddles and strangles have similar profit profiles (see Figures 7.3 and 7.4). The choice between the two would be largely a matter of price.

Either of these would be sold by someone looking for a decrease in volatility, probably caused by a sideways moving market. Volatility will normally decline in a flat market although an overhanging threat may lead to an increase in demand for options and a consequent increase in volatility.

Bull call spreads and bear put spreads are also commonly traded. The former is the buying of a call option and the selling of a call option with a higher strike price. It will always involve the net payment of a premium because the option being bought has a strike price closer to the market. The spread is bought when a trader is bullish for a limited move. It is also a useful way of subsidising the cost of the nearer option, particularly when high volatility makes option premiums appear expensive. By buying one and selling the other the trader is limiting profit to the

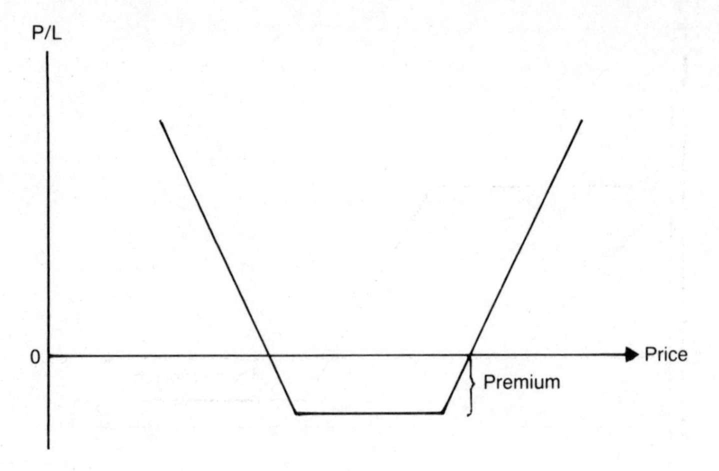

Figure 7.4 Strangle.

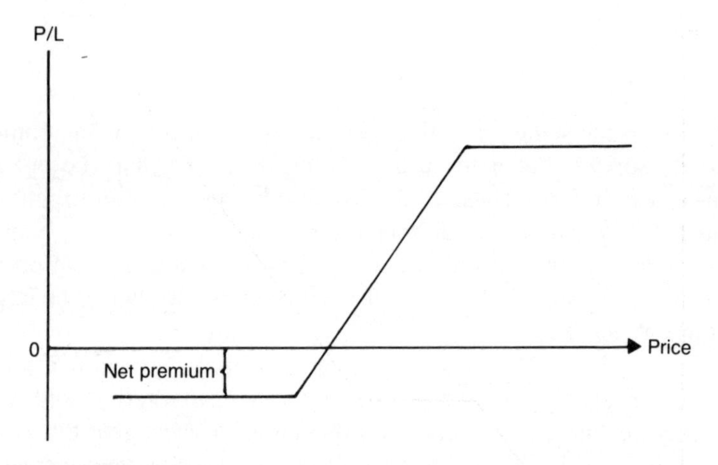

Figure 7.5 Call spread.

difference in strike price less the premium (see Figure 7.5), but in return for giving up potential profitability is not paying for the volatility and time value (because the two call options will have the same time value and virtually identical volatility).

A bear put spread is the opposite – the buying of a put option and the selling of a put option with a lower strike price. The maximum profit of the difference in strike less the net premium is achieved if the market is exactly on the lower strike at expiry (see Figure 7.6).

Another way of reducing the cost of an option is to buy one option and sell the other kind, that is to buy a call and sell a put or buy a put

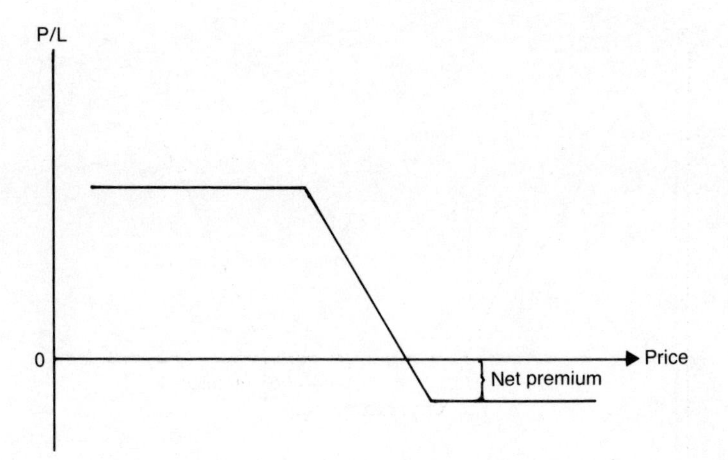

Figure 7.6 Put spread.

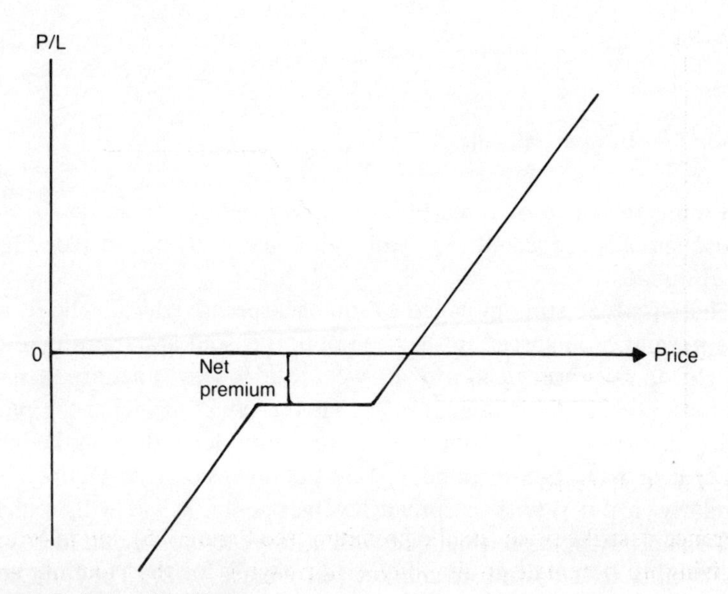

Figure 7.7 Buy call, sell put.

and sell a call. These have different profit/loss profiles (see Figures 7.7 and 7.8) because the trader is not giving up profitability if the market moves in the direction of his bought option, but loses money if he is wrong about market direction.

Sometimes more options are sold than bought, thus reducing the cost of the bought option but increasing the risk. Usually a trader would not

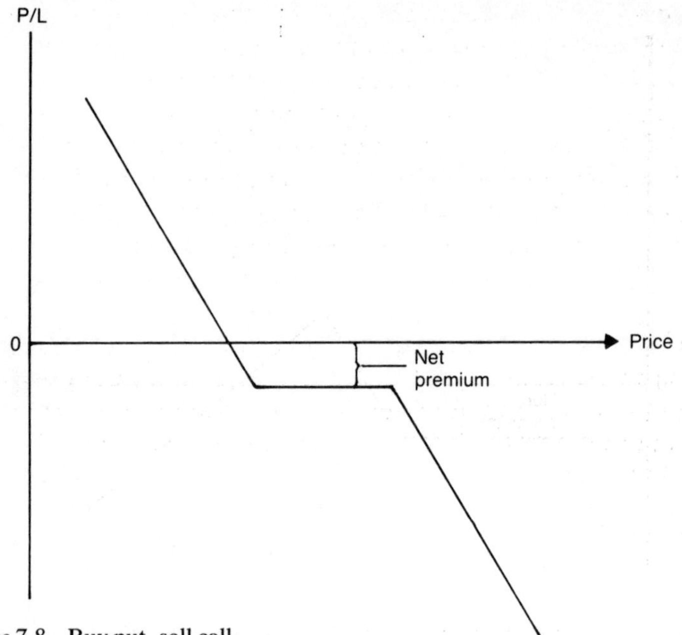

Figure 7.8 Buy put, sell call.

sell more than two or three times as many options as he has bought. These spreads are called ratio call (see Figure 7.9) or put (see Figure 7.10) spreads.

There is also a strategy called a ratio backspread, which is the reverse of a normal ratio spread in that one option is sold and more than one bought. In this case, a call with a lower strike is sold and more than one with a higher strike is bought. This has the effect of increasing profitability in a rising market, limited loss (or profit depending on the initial net cost or collection of premium) on the downside (see Figure 7.11). Similarly, a put ratio backspread involves selling a put with a higher strike and buying more than one with a lower strike to gain increasing profitability in a downward market (see Figure 7.12). They might be used when a buyer is looking for increasing activity and higher volatility but is more confident of the direction of that move than the buyer of a straddle.

A calendar spread is the buying of an option in one month and selling one, usually with the same strike, in a further month. The effect of volatility and time value on the net premium is different from when both options are in the same month because volatility in the two months may be different and time value will be declining faster in the nearer month.

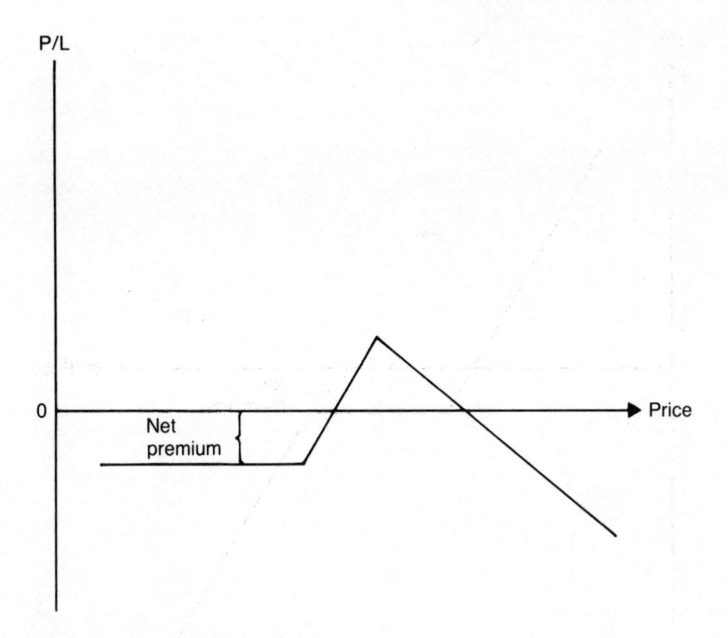

Figure 7.9 Ratio call spread.

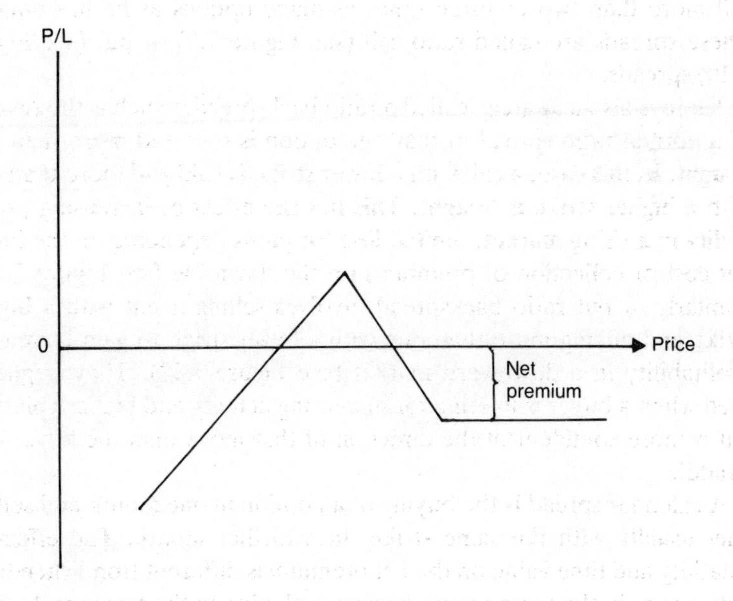

Figure 7.10 Ratio put spread.

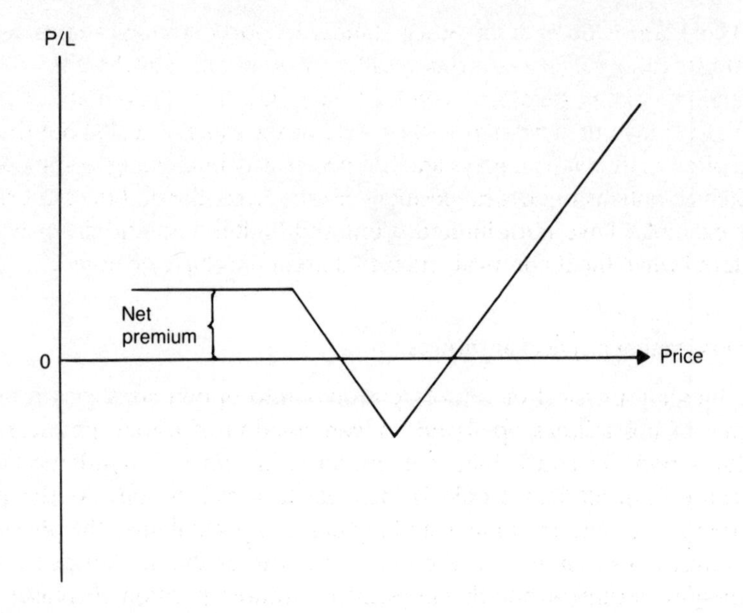

Figure 7.11 Ratio call backspread.

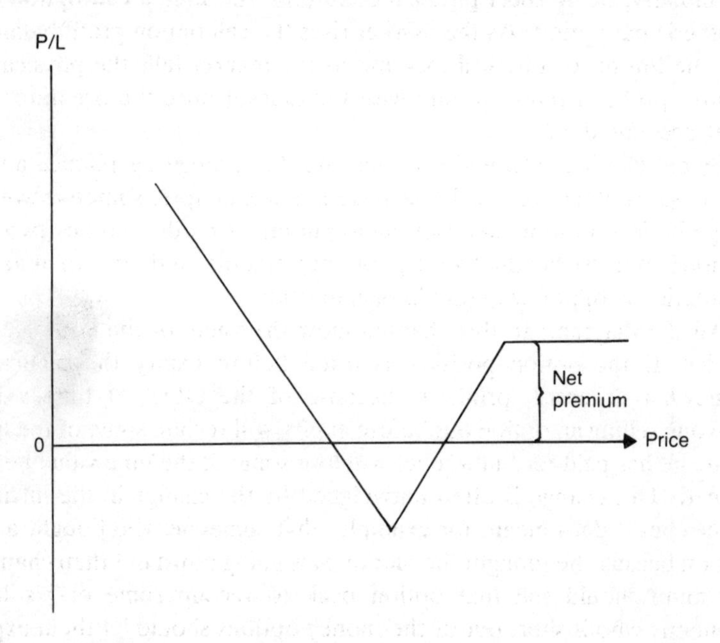

Figure 7.12 Ratio put backspread.

There are a number of other standard option strategies such as a butterfly (long call at x, short two calls at y, long call at z where $x<y<z$ or the reverse for puts) or a condor (long put at v, short put at x, short put at y, long put at z where $v<x<y<z$ or the same for calls) but these are often mathematical plays and not necessarily interesting to someone matching options to physical positions. Both the condor and the butterfly, for example, have both limited profit and limited loss and are usually entered when the theoretical and actual premiums have diverged.

Incorporating physical or futures

By building physical or futures positions into option strategies a new range of possibilities opens up. A combination of a long physical or futures position and a long put option is known as a synthetic call, because its profit/loss profile is identical to a call option. As the put option moves into the money and increases in profitability, the physical or futures position loses at the same rate. But as the market rises, the option is worthless but the physical or futures position increases in value dollar for dollar with the market (after the premium has been recouped).

Similarly, being short physical or futures and long a call option is a synthetic put option. As the market rises the call option profit balances out the futures or physical loss and as the market falls the physical or futures profit increases in line with the market once the premium has been accounted for.

By putting these synthetic options into the strategies explained above the desired objective can be achieved. For example, someone with a long physical position and wanting to put on a straddle will buy two put options, one to be the put leg of the straddle and one to make a synthetic call option with his physical position.

All the diagrams in this chapter show the value of the strategies at expiry. If the option position is lifted before expiry the picture is somewhat different, primarily because of the effect of time value. Anyone selling an option out before expiry will recoup some of the time value he has paid for but a seller will lose some of the time value he has gained. This change is often outweighed by the change in the intrinsic value, but it does mean, for example, that someone who bought a put option because he thought the market was going down but then changed his mind should sell that option back to recoup some of his loss. Someone who is short out-of-the-money options should let them expire rather than buy them back.

Delta hedging

Delta hedging is a refinement of straightforward hedging using options. Delta is defined as the rate of change in the option premium with a change in the underlying futures price. An at-the-money option has a delta of 0.5 which means that the premium will increase or decrease at 0.5 of the change in the underlying futures price. A heavily in-the-money option has a delta approaching 1.0, so that any change in the underlying futures price is reflected almost cent for cent in the premium. Conversely a heavily out-of-the-money option has a delta approaching zero, meaning that a change in the underlying futures price has virtually no effect on the premium. An instantaneous reading of delta is given by all the on-line and other computer pricing models for options.

When using options to hedge it is usually desirable to achieve an equivalent options profit to any physical loss. An equal volume of options and physical barrels will not give this effect because the option premium will not change at the same rate as the futures or physical price. It is therefore necessary to adjust the number of options to get the correct profit. If, for example, at-the-money options are being used, it will be necessary to have twice as many option barrels as actual barrels because the option premium will only move half as fast as the underlying futures.

But any price move will put the option in or out of the money and will therefore alter the delta and therefore the number of options necessary. As the market price, and therefore delta, changes, it is necessary to adjust the number of options at frequent intervals to maintain the correct cover. In practice most traders will look at the position, and make any necessary changes after certain price movements.

Example

A trader holding 50,000 barrels of crude oil wants to hedge using at-the-money options on the Brent market. The current market price is $19.00 per barrel. His requirement is for 100 put options on day 1.

	Market price	Delta	Options required
Day 1	19.00	0.50	100
Day 2	19.50	0.45	111
Day 3	20.00	0.40	125
Day 4	19.00	0.50	100
Day 5	18.00	0.70	71

8

How the oil industry can use the futures markets

In this chapter we examine some of the different ways in which the oil industry can use the futures markets to complement its physical trading activities. In fact each company using the markets develops its own preferred trading methods from these basic theories and will analyse its risks and consequent futures requirements in a slightly different way, just as they do on the physical markets.

In all the examples given in this chapter it is assumed that hedges and other futures positions are left in place until the related physical position has been disposed of. In practice, most futures positions are not left in place but are lifted and replaced, in whole or in part, as the user's opinion of the market changes. A perfect hedge applied throughout a physical position's life would indeed virtually eliminate risk but it would also remove the opportunity for increasing profit. Most traders wish to take advantage of price moves in their favour while gaining protection against an adverse move. Futures and options markets should make it possible to do this, but they do not remove the need for an opinion of the market.

Supply department

Supply departments in any large company have, by their very nature, an unpredictable life. They can find themselves unexpectedly long or short of product, or facing sudden changes in stocks or production levels. As they constantly try to balance these supply and demand fluctuations, the futures and options markets can be used to overcome some of the problems. One of the main advantages of futures and options trading is the ease with which positions can be adjusted or removed altogether.

This ease is partly the high level of liquidity and partly the administrative ease of handling futures. For example, if a position is taken on the futures markets and then removed, the position can be closed out and the deal is closed. If the same protection is taken on the cash market and later closed, the department may still have to become involved in nominations, book outs and other administrative details even if the cargo deal was reversed some weeks previously.

Judicious use of the futures market can also substantially reduce the cost of financing oil stocks. Instead of holding stocks against known future demand, the stock can be sold and the demand requirements met by buying futures, freeing up both capital and storage. Conversely, occasionally product can be bought from the market one month and sold back the following month at a premium sufficient to cover the costs of storage and delivery. It is rarely possible to make much profit on this trade, but it can be used to make better use of spare storage capacity.

Trading the cash and carry, as this is called, should be prefaced by careful calculation of the costs involved as interest rate changes and, in Europe, currency fluctuations can eliminate the potential profitability of the trade. If a cash and carry is being done some way forward, currency and interest rates should be fixed at the time the deal is done.

Example

In March, Company A has gas oil in tank against early May commitments. At the time, March gas oil is trading at $146.50 and April gas oil at $142.00. So the company sells March gas oil and buys April, allowing the contracts to run through to delivery. The timing of the delivery is at the buyer's option, although there are circumstances where the seller can reject the buyer's chosen delivery range. It has to be assumed, however, that the buyer will elect the last possible moment, 31 March, to take delivery and the seller will not therefore be paid until 5 April (five days after bill of lading). When the company is taking delivery during April it will choose the latest possible delivery range but may be forced to change to 21–25 April for its delivery. It would then send its barge on 25 April, paying on 30 April.

The total cost of the transaction can be calculated as follows:

	$/tonne
Payment received 5 April	146.50
Interest at 8% for 25 days	0.81
	147.31
Payment made on 30 April	142.00
Commission	2.00
	144.00

Although there are the additional costs of delivery arising from the physical movement of the product, these have to be calculated on a case-by-case basis because different factors come in to play. For example, the product to be delivered may already be in an acceptable installation or it may have to be moved; storage freed up for most of April may or may not give a financial advantage.

Similar calculations can be made in reverse for the cash and carry. When calculating the cash and carry it should, however, be remembered that arrangements must be made for leaving the product in the storage installation. Dutch law prohibits the holding of third-party material in a refinery, so anyone receiving delivery at one of the refineries in the region would be obliged to move it to one of the independent storage installations.

A supply department may also wish to trade options against any product it may have in store. By selling call options it can generate income from its stocks. Although it will effectively be putting a maximum value on the stock, it will have the necessary cover in an upward price movement. It would only sell options if it felt the market was likely to move lower or remain more or less unchanged. In a flat market it would stand to gain not only from decreasing time value (which always works to the seller's advantage) but also from the decreasing volatility.

Example

In early 1989, prices had risen steadily since the OPEC meeting but began to show signs of instability around April. A supply department, required to maintain its physical gas-oil stock levels, might have wished to gain an income stream by selling gas-oil calls with a strike price slightly above the market. With the market trading at $155 in mid-April, the department might have sold $160 call options for July at $5.80 per tonne. Prices drifted lower over the next few weeks and the option would not have been exercised on expiry, when July futures were trading around $143.

The supply department would therefore have received an income of $5.80 per tonne (less commission and financing costs) to offset against the drop in value of its stocks. Had prices risen through $160, it could have bought back the option or used the short position effectively created to hedge its stocks, depending on its view of the market.

The income generated would not have been equal to the drop in value of the stocks unless the department had chosen to delta hedge, in which case the options would have been bought and sold repeatedly to maintain full dollar-for-dollar cover (see Chapter 7).

If they had chosen this route, the sequence of events could have looked as follows (for each 100 tonnes in stock):

	Delta	Options required	Number bought or sold
Day 1	0.6	1.67	1.67
Day 2	0.55	1.82	0.15
Day 3	0.65	1.54	0.28

The frequency of adjusting the position can be determined by the user to maximise protection without taking all his time and costing too much in commission. Some people choose to adjust their positions every day and others once a week or after a certain move in the delta.

A supply department might also use futures to take cover against a change in its physical position.

Example

Company B's supply department receives a call from its refinery to say that production problems have meant a shortfall in gasoline availability. The supply department had already sold the material and will now have to buy gasoline on the open market. But it fears that prices are rising and will rise further before it is able to find suitable material.

So, while it is looking for suitable gasoline to buy, Company B goes onto the futures market and buys the equivalent amount of futures. When it finds the physical material it will sell out its futures contracts (or enter into an e.f.p. transaction).

Futures		Physical	
	cents/gallon		cents/gallon
Buy futures	69.45	(Trading at	67.65)
Sell futures	70.30	Bought at	68.90
Profit	0.85		

The futures profit can then be offset against the physical price to give a net price of 68.05 cents/gallon. Although this price is a little higher than the physical price when the futures were bought, this has resulted from a change in the differential between the appropriate physical grade and the futures market, known as the basis.

A supply department might also choose to sell some or all of its material on an e.f.p. basis. Using an e.f.p. enables any futures market user to separate the pricing of a contract from the physical supply of the product or crude.

Example
The supply department of Company C wishes to re-establish term supply contracts with Company D but does not want to enter into fixed price or variable pricing arrangements based on published price information. Using an e.f.p., it agrees to sell Company D 4.0 million US gallons of heating oil per month, prices to be fixed on the basis of the NYMEX heating-oil contract prices for that month. All the physical contract details are agreed as normal and it is agreed that an e.f.p. for 95 lots of heating oil will be registered on the last trading day for each month. The e.f.p. will be registered at the previous night's settlement.

Each month, Company C will decide when it wishes to price its sales. It has agreed a premium of 25 points to the NYMEX for locational reasons. Its net sales price could be calculated as follows:

Futures		Physical	
	cents/gallon		cents/gallon
Sold at	51.00		
E.f.p. registered	49.00	Company D invoiced	49.25
Futures profit	2.00		

The net sales price is therefore 51.25 cents/gallon, equivalent to the 51.00 cents/gallon futures sales price plus the 25 point premium.

The price at which an e.f.p. is registered is irrelevant because it is also the price at which the physical deal is invoiced. In the above example, if the e.f.p. had been registered at 55.00 cents/gallon, the futures loss would have been 4.00 cents/gallon, but the physical sales price would have been 55.25 cents/gallon and the net price would still have been 51.25 cents/gallon.

Using an e.f.p., both buyer and seller can choose the time to fix the price. The two prices are independent and unknown to the other party.

The producer

Crude-oil producers are always worried about falling prices. They can use the futures and options markets as a straightforward hedging medium or to fix a minimum price. If a producer is convinced that the market is going to fall, he can simply sell futures. Then, if he is right, he will have a futures profit to add to the lower income he will receive for his oil.

If he is wrong, however, and the market rises, he will lose out on the upside benefit for as long as his futures position is in place. By buying options, he will be able to participate in any upward move while obtaining protection against a downward one. As with anyone buying options, he needs to consider what he is trying to achieve and how much he is prepared to spend to achieve it. Much of this decision will depend on his view of the market. If he is bearish for a major move, he may like to just buy puts or buy puts and sell calls to reduce the cost. But if he is bearish for a limited move he may buy a bear put spread (see Chapter 7) to reduce his outlay by selling a lower strike put than the one he is buying.

Producers are often sellers of call options as well as buyers of puts. If they feel the market is overvalued they can sell call options either at-the-money or just out-of-the-money. This way they receive an income from the premium. This does not, however, provide any protection against a downward move and in an upward move the effect is to fix a maximum selling price at the strike price plus the premium while the option positions are still open.

The e.f.p. is also a very useful tool for an oil producer. He can agree a contract with his customer and then use the futures market to price the deal. The differential between the crude he is selling and the crude traded on the futures market will have to be agreed, as will the date of registration. By using the e.f.p. mechanism, a producer can agree a term supply contract with each delivery priced using the futures market.

Example

Producer A agrees to sell a cargo of 450,000 barrels of Ninian crude to Refiner B each month for the next year. The cargoes will be priced on the following month's contract on the IPE Brent contract (because the Brent contracts expire on the 10th of the previous month) and the e.f.p. will be registered on the bill of lading day or the last trading day of the contract, whichever is first, at the previous day's settlement. The differential will be the five-day average around bill of lading day as published by Screen Services. Using the March delivery as an example and assuming the e.f.p. will be registered at $17.40 and the differential is 10 cents discount:

Producer A		Refiner B	
	$/barrel		$/barrel
Sells 450 Brent at	17.75	Buys 450 Brent at	16.80
E.f.p. registered at	17.40	E.f.p. registered at	17.40
Futures profit	0.35	Futures profit	0.60
Invoices B	17.30	Invoiced by A	17.30
Futures profit	0.35	Futures profit	0.60
Net price	17.65	Net price	16.70

In both cases, therefore, it can be seen that the net price is the price of the original futures contracts less the differential.

Producers may also like to use trigger pricing. In this case the producer will agree with the buyer to sell a cargo of crude, to be priced at either buyer's or seller's option over a fixed period. Normally a minimum size for each trigger will be determined. Then, if it is at buyer's option, the buyer will call the seller when he wishes to price the cargo in part or in whole. At this point the price will be fixed and the seller must take the appropriate action. The seller may already be hedged by having sold futures. He would then have to lift that part of the hedge which has just been triggered. If he is not hedged, because he thinks prices will rise, he may wish to take up a long position when the buyer prices. If it is at seller's option, the producer will call his customer when he wishes to price some of the contract.

The refiner

Refiners have had to face constant price fluctuations in the price of both crude and product contracts in recent years, making any planning of the future costs very difficult. Although refineries have returned to profit in the latter part of the 1980s, refiners still remember the long years of loss-making and are using the futures market to remove some of the uncertainties and make realistic estimates of refinery netbacks.

Many refiners still calculate the viability of running crude by using today's crude and product prices. Although this may even out overall, the picture can alter between the date of purchase of the crude and the sale of the resultant products. And anyone contemplating a processing deal needs to link the product prices directly with the products produced.

Futures markets can be used to lock in a future crude purchase price, a product selling price or both. Perhaps the most attractive of these, in normal market conditions, is the selling of products forward to guarantee a return on the barrel. On NYMEX a large proportion of the barrel can be hedged in this way because there are contracts in all the major products, gasoline, heating oil and fuel oil, although the last has yet to establish itself. In Europe the only products available at present are gas oil and the new fuel-oil contract and in Singapore fuel oil is the only contract.

A refiner looking to hedge must then decide whether he wishes to hedge some of his other products in a different market place. The NYMEX gasoline market, for example, can provide protection against major adverse price moves in Europe, but it is a completely different market reacting to different influences. In the summers of 1988 and 1989, for example, the US gasoline market was very strong and moved far ahead of the European market, making it a very unsatisfactory hedging vehicle.

Example
In January a refiner buys 100,000 barrels of WTI for $19.80 per barrel. He is worried that prices will fall by early March when he will be looking to sell the products on the spot market. As a US refiner, he is able to use both the gasoline and the heating oil markets on NYMEX and therefore hedge around 70 per cent of his output.

January
Buys WTI for $19.80 per barrel
Sells gasoline futures at 61.00 cents/gallon
Sells heating-oil futures at 51.40 cents/gallon
Sells fuel-oil futures at $17.00 per barrel.

March
Buys gasoline futures at 60.00 cents/gallon (1.00 cent/gallon profit)
Sells physical gasoline at 59.90 cents/gallon
Buys heating-oil futures at 51.00 cents/gallon (0.40 cent/gallon profit)
Sells physical heating oil at 50.90 cents/gallon
Buys fuel-oil futures at $16.50 per barrel
Sells physical fuel oil at $16.60 per barrel

This gives him an effective gasoline cash price of 60.90 cents/gallon and an effective heating oil price of 51.30 cents/gallon and an effective fuel oil cash price of $17.10 per barrel.

This type of hedging does not guarantee the refiner the price he originally sells the futures contracts because prices in the forward months do not always reflect the moves in the spot month exactly (until the forward month becomes spot). The forward month may be at a premium or discount to the physical market when the hedge is originally put on. The refiner should be careful when selling forward that the price differential is realistic. In a forward discount or backwardation market the spot month is at a premium and selling the forward months may not be an attractive proposition.

The refiner might decide not to sell the products forward, perhaps because he thinks the market will rally. But he might want to take some cover. He could buy put options in the product markets, so that he could take advantage of any upward move in the market but be protected against a downward move.

For example, with prices similar to those in the above example, the refiner might decide that instead of selling the gasoline and heating oil markets he will buy 60 cent gasoline puts and 50 cent heating-oil puts. Then, if the market falls below these levels he will exercise his options (or take the profit by selling the options out) but if it rallies he will abandon the options.

He might also decide that he will sell call options because he thinks the market will be generally stable and option volatility will decline. He would then receive the option premium and sell his physical product at the same or similar price to that when he bought the crude. If, however, the market rallies, he will have to buy the options back, probably at a loss, or use his physical material as cover in which case he would have set himself a maximum selling price of the strike price plus the premium he received.

The refiner can also use the crack spread to fix both the purchase price of his crude oil and the selling price of his products. The crack spread is the buying of crude and the selling of an equal number of product contracts or vice versa. In one pit on the NYMEX floor the crack spreads (either three crude, two gasoline, one heating oil or five crude, three gasoline and two heating oil) are quoted as differentials. The actual prices are fixed later. As with all spread trading the actual price levels are unimportant, it is only the differential that is important. A typical crack spread using fuel oil has yet to be established.

Some people prefer to build their own crack spreads, using crude for one month and products for the next month for example, but these do not attract the beneficial original margins. It is now possible, too, to trade a transatlantic crack spread with the crude in one market and the products in another. This is particularly attractive for US refiners importing North Sea crude into the United States and selling the products on the local market.

Example

A refiner wants to hedge his Brent purchases and product sales ahead because he can see an attractive refinery margin built in to the futures prices. So, during February, he buys June Brent and sells July heating oil, gasoline and fuel oil. When the crude purchase is made physically he sells his Brent futures (or simply lets the contract expire because Brent has cash settlement not physical delivery). Then, when he sells his physical product, he buys back his product futures contracts. He will sell three gasoline, two heating-oil and one fuel-oil contracts for every 6 crude contracts he buys as this most closely matches his product slate.

Futures		Physical	
Brent			
6 bought at	$17.40 per barrel		
Gasoline			
3 sold at	70.00 cents/gallon		
Heating oil			
2 sold at	54.60 cents/gallon		
Fuel oil			
1 sold at	$17.80 per barrel		
(paper refining margin	$7.90 per barrel)		
Brent			
6 sold at	$16.00 per barrel	Bought at $16.00	
Gasoline			
3 bought at	64.00 cents/gallon	Sold at	63.75 cents/gallon
Heating oil			
2 bought at	50.00 cents/gallon	Sold at	50.20 cents/gallon
Fuel oil			
1 bought at	$16.00 per barrel	Sold at	$16.05 per barrel
(paper refining margin	$7.11 per barrel)	(margin	$7.09 per barrel)

By executing this trade the refiner has given himself a crack spread profit of $0.79 per barrel to give a total margin of $7.88 per barrel, virtually the same as that he originally fixed, although the actual margin has fallen to $7.09 per barrel.

As with all other uses of the market it can be done in similar fashion using options, though the buying of a crack spread in options can be expensive in premiums.

A refiner can also use e.f.p.s both to buy crude and sell products and can enter trigger pricing deals.

The traders

Traders were the first sector of the oil industry to trade the futures markets actively and continue to be a major part of the market. Some use the futures markets to add to or as an alternative to their physical positions rather than as a hedging vehicle, transferring some of their speculative trading away from the physical market because of the administrative ease of futures with no shipping and locational problems.

Looking at their use of the futures markets as a complement to their physical activities, however, hedging is probably the largest use, as with other sectors of the industry. If a trader holds a long position, for example, and fears that prices are falling, he can sell futures as an alternative to selling the physical cargo.

With the physical market being very transparent, other traders often know the placing of cargoes and know that a particular trader has a need to sell or buy. The trader can then find that he is unable to find another party wishing to trade at an acceptable price. But if he is hedged on the futures market he can afford to wait longer before making a physical deal. The cargo need not become distressed. The confidentiality of the futures market can be a major attraction in a case like this. Later he may decide that rather than lift the hedge he will trade the physical cargo on an e.f.p. basis and use the hedge position as his futures position for the e.f.p.

Again, the trader has to decide which futures market to use. If he has a strong opinion about the relative strength of one market over another, he may decide not to hedge in the nearest equivalent market but to trade a different one and try and take advantage of the differential as well as hedge his cargo. This will probably make a subsequent e.f.p. impossible, but may have some attraction, particularly if the cargo under consideration does not have an equivalent futures market. For example, someone with a European gasoline cargo might well decide that Brent provides a better hedge than the NYMEX gasoline market and will therefore use Brent.

Example

A trader has a cargo of gasoline in North-West Europe and is concerned about falling prices. He cannot find a satisfactory buyer, so he decides to sell futures. He looks at the NYMEX gasoline market, but decides that this market is much stronger than the European market and it will not therefore provide him with much protection against falling prices. His cargo is not suitable for export to the United States so the option of sending it across the Atlantic is not there. So he decides to sell an equivalent quantity of Brent futures.

When he finds a buyer for the physical cargo, he may try to do an e.f.p. against the Brent or he may, more probably, simply lift the hedge.

Futures	Physical
Sold Brent $17.20 per barrel (equivalent price $143.28 per tonne)	Long gasoline $278.00 per tonne
Bought Brent $16.70 per barrel (equivalent price $139.16 per tonne)	Sold gasoline $272.00 per tonne

He has therefore made a profit of $4.12 per tonne to offset against the $6.00 loss he has made on the physical market. So although the hedge was not perfect, it has provided some cover in the falling market.

Even if there is a futures market in the product or crude the trader wants to trade, the relationship between the futures price and the physical price will vary from time to time. As expiry approaches, the two prices must converge if the market is successful, but at other times there is a differential known as the basis. A trader hedging on the futures market is assuming basis risk, that is the risk that the differential will vary and the degree of cover provided will therefore change.

The differential between the futures and physical price for the same commodity can provide a useful trading opportunity. The trader can trade this differential in the same way as any other: buying the 'under-priced' product and selling the other. For example, he might decide to buy futures and sell the physical product because the physical product prices are higher. These opportunities are usually short-lived and the profits are unlikely to be large.

It is important to be certain that the difference in price does not reflect an important difference in the two products. For example, the physical gas-oil price may be higher in the short term because of bad weather preventing Russian gas oil getting into North-West Europe. If this delay is likely to continue into the delivery period of the exchange, the futures price will reflect the problem, if not, then the physical price for immediate delivery will increase but futures will not. This means there is a difference between the two products and that difference is reflected in the price. This would not give a good trading opportunity to most traders.

Inter-month spread trading can also provide useful opportunities. Although the cash and carry (buying one month and selling at a later one, settling out the trades by taking delivery one month and making it the next) is eagerly watched for and profits are limited, it can produce some profit particularly for those with access to cheap storage or who have storage unused. Similarly, lending existing stock to the market by delivering one month and taking back the next can be attractive in a backwardation market.

The most important point to check in either of these trades is the exact delivery mechanism as this can make all the difference between profit and loss. In particular, delivery timing is usually (except on SIMEX) at buyer's option so it cannot be assumed that a buyer will take product at any one time.

Another switch which the traders watch for is the arbitrage between NYMEX and IPE. The term is usually taken to mean the spread between gas oil and heating oil. The relationship between the two varies enormously depending on weather conditions, local production, and consumption levels and sentiment. Over the last five years, heating oil has normally been at a premium to gas oil, but there have been occasions when particularly cold weather in Europe, for example, has reversed the premium.

When doing the arbitrage the trader must decide what conversion factor he wishes to use to convert the heating-oil price to dollars per tonne or the gas-oil price to cents per gallon. The exact differential is not important provided that the same one is used throughout the trade and for assessing past performance. Probably the most popular conversion is 313 gallons per tonne (the conversion at the standard gas-oil density of 0.845 kg/litre) though 310 or 309.54 are also popular.

It is also necessary to trade the same quantity of product on each market, which means trading four gas-oil contracts to each three heating-oil contracts because of the difference in size of the contracts.

Some traders also like to trade a later month in one market than the other in order to preserve the theory behind the arbitrage that product can be lifted from one market and delivered on to the other. Thus when opening the arbitrage, the selling would be done in the month after the buying.

Example
The February arbitrage has been trading in a $3.00–9.00 range and a trader believes that the potential for moving out of this range is small. When it comes back towards the lower end of the range, therefore, he decides to buy heating oil and sell gas oil. As it widens out again he then takes the position off as it approaches the top of the range. In this case, there is no question of taking delivery of either market so the fact that the arbitrage does not reflect the cost of transport is not important.

IPE			NYMEX		
	$/ tonne	cents/ gallon		cents/ gallon	$/ tonne
4 gas oil sold	157.00	50.16	3 heat bought	51.20	160.25
4 gas oil bought	163.00	52.08	3 heat sold	54.65	171.05
Loss	6.00	1.92	Profit	3.45	10.80

The net profit on the trade is therefore $4.80 per tonne or 1.53 cents/gallon, less the cost of trading.

One point worth noting is that almost all the sharp movements of the arbitrage have seen a rapid increase in the NYMEX premium rather than a rapid decrease. This is largely because the NYMEX market tends to react more strongly, in the short term at least, to changes in sentiment.

The arbitrage can be done using options rather than outright trading though this tends to be a rather expensive exercise.

Many traders get involved in processing deals and other aspects of refining. They will therefore use the various trading methods open to refiners, particularly crack spreads. More recently, many traders have been developing upstream or downstream activities and moving away from some of the more openly speculative areas that formerly constituted their main activity. They may therefore become involved in any trading methods applicable to those areas.

Traders can also use options, either as an alternative to futures or as a means of speculation. They can use any of the options strategies outlined in Chapter 7 and may be particularly interested in some of the volatility strategies.

The marketing department

The main aim of the marketing department of a major company is to expand its company's market share while maintaining profitability. The futures market can help such a department lock into profit margins over a longer time and offer fixed prices on to its customers. A number of customers, particularly large public bodies and manufacturers, are attracted by fixed prices so that they can work within their budget.

Of course, the fixed price is unlikely to be the best price, when the time comes, for both parties. It will either be below the price on the market at the time or above it. But the advantage comes from knowing what it is going to be in advance. And over a long period it is likely that an average price will be achieved by both sides. This may not, however, be seen for some time and it is difficult for any purchaser to explain why, despite keeping within his budget, he has paid above the market price. There is therefore always a degree of compromise about fixing prices forward. Consumers were slow to come into the oil futures market because, for the first few years they were around, prices tended to fall. That situation has been reversed more recently and consumers are becoming more interested, leading to a more mature and satisfactory market.

It is necessary for marketing departments outside the United States, who frequently operate in local currency, to fix their currency exchange rates at the same time as the oil price.

The simplest method is to buy futures and use that price as the basis for selling forward. This is effectively a simple hedge.

Example

In April, Company A buys gas-oil futures at $154.00 per tonne for November. It then sells the product on to its buyer at a price based on this (fixing its currency at the same time). Come November, the price of gas oil has risen to $173.00 per tonne for November futures and the marketing department is charged the current physical price of $172.00 per tonne by its refiner.

Futures		Physical	
	$/tonne		$/tonne
Buys November	154.00	Sells to customer	158.00
Sells November	173.00	Buys from refinery	172.00
Profit	19.00	Loss	14.00

The marketing department has achieved a net profit of $5.00 per tonne on the deal, comparing favourably with the $4.00 per tonne built into the price passed on to its customer. (An allowance for the fluctuations in the differential between futures and physical can be built into the selling price.)

Similar deals can be arranged for any product with a futures market.

Rather than actually buy futures, the marketing department might decide to use options, particularly if it thinks the price might fall. Using the same example, the department might decide to buy November call options with a $155.00 strike. This would probably cost it several dollars, because it is at the money and some way forward. It then becomes a matter of trying to balance out the risk reward ratio of the transaction.

If the department did decide to take the option route, it would only exercise the option or sell it out if the price was above $155.00. If it was below, it would simply buy the product from its refinery at the market price and sell it on to its customer at the agreed price.

The distributor and large consumer

Both distributors and large consumers may wish to fix their buying prices in advance by buying futures against their planned requirements. As with the marketing department, it is necessary to fix the currency exchange at the same time if appropriate. Some distributors have found such forward price fixing, passed on to their customers, a good marketing tool.

Unless consumers are very large, they may find that entering the futures market themselves is unnecessary but they can achieve the same protection from their supplier. In most inland markets, prices tend to lag behind those on the spot market, and also therefore the futures market, particularly when prices are falling. For this reason, futures are not a very good hedge for those using inland prices for their purchases.

Futures are not a good way of buying forward for those in countries where the selling price of oil products is fixed by the government.

For those consumers and distributors who buy their products on a spot-related basis, however, futures can be a useful tool. When a consumer or distributor finds the price of the oil product attractive, either because he fears an upward move or because the price is at or below the budgeted price and he wishes to guarantee some of his purchases at that level, he will simply buy futures. Then, when the time comes to take physical delivery from their supplier, he will sell the futures. If prices have risen, he will have a futures profit to offset against the higher price of the physical oil.

Instead of buying futures, the consumer or distributor may like to look at buying options to protect against a major upward move in prices but allow it to benefit from any fall in values. In this case, the company would buy call options with a strike price at the appropriate level. When looking at option purchases in this way it is necessary to establish what cost the company is prepared to pay for what degree of cover.

Some companies prefer to allocate a fixed sum of money while others prefer to match volumes. Some prefer to buy at-the-money options, while others prefer to buy a larger number of out-of-the-money options, or the same number at a lower cost. If the market rises the value of out-of-the-money call options will also rise, though not as fast as the nearby ones (see Chapter 7). But out-of-the-money options will often have a lower implied volatility and therefore can be seen as 'better value'. As with most things, the user has to make his choice. Buying call options will enable the consumer or distributor to fix his maximum buying price at the strike price plus the premium paid.

All these ways of using the market are mirrored by services provided by the Wall Street refiners and over-the-counter options traders. It is up to the user whether he wishes to assume risk himself, lay it off in whole or in part using the futures and options markets, or lay it off entirely and pay someone else to take it on.

9

Charts and chartists

The term 'chartist' is used to describe people who attempt, often successfully, to predict future prices by analysing past price movements and identifying trends and patterns. The oil markets have, during their short history, proved themselves attractive to these commodity chartists. There are as many ways of analysing charts as there are chartists, but it is useful to look at a few standard techniques on which most chart systems are based.

When the energy futures markets first began trading the oil industry was, inevitably, highly sceptical about charts. Why should a market react in a particular way just because the price movements had made a certain formation over the past few weeks? Now, however, most of the successful users of the markets have a grudging respect for the charts and *do* take account of what the charts are saying.

Charting theory is based on the principle that all that is known about a commodity is reflected in its price. It is therefore possible to trade different markets by using charting theory and having no knowledge of the fundamentals of the commodity. Technical trading systems are based on pure analysis of these prices.

Charts are often thought to be a self-fulfilling prophecy. If everyone thinks that selling will come in at a particular level, that is the level potential sellers will pick and the chart point will be given validity. But of course if markets always obeyed technical analysis theory prices would never move through chart points. Fundamental news will take a market through a level which would be expected to provide support or resistance.

An oil-industry user of the market will probably not use charts to determine what to do in the market, but may well change the timing of his action because of what the charts are saying.

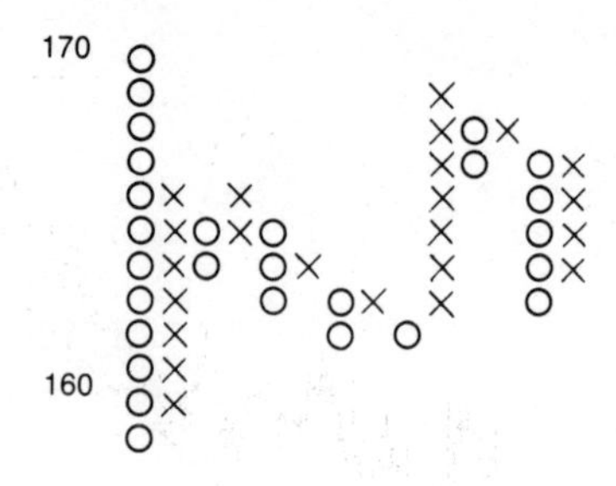

Figure 9.1 A point-and-figure chart.

In general, the US markets react more strongly to chart indications than the London markets, largely because of the higher speculative element. Large amounts of individual speculators' and commodity-fund money is tied up in chart systems and great attention is paid to sell- and buy-stops and support and resistance levels.

All charts are based on the principle of recording price movements over a period of time. They usually involve the plotting of every move a market makes (a point-and-figure chart) or the daily range and closing price (a line-and-bar chart). Often the volume, open interest and, perhaps, relative strength index (a variable worked out on a formula involving price movements over a period of time) are also plotted.

Point-and-figure charts

A point-and-figure chart (see Figure 9.1) involves the recording of an 'O' for every downward move of a certain size and an '×' for every upward move. The size of the move is variable, but would normally be, say, one dollar per tonne on the IPE gas-oil contract and 0.25 cents/gallon on the NYMEX gasoline and heating-oil contracts.

Once the size of the movement has been chosen, the points are plotted on the graph each time the market moves. For example, heating-oil moves on the New York market using a 0.25 cents/gallon movement would be plotted as follows:

Sequence of trades	Action taken
78.00	starting point
78.10	none
78.30	'×' plotted

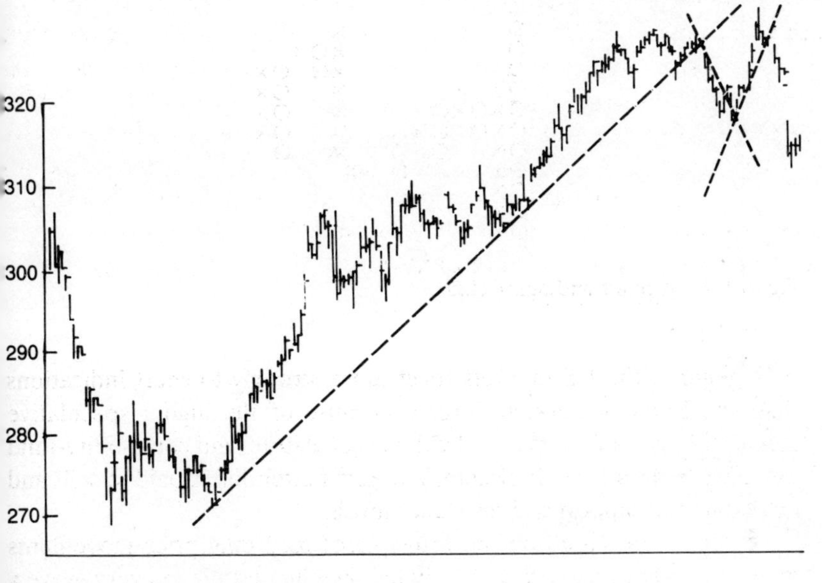

Figure 9.2 A line-and-bar chart.

Line-and-bar charts

A line-and-bar chart (see Figure 9.2) entails the plotting of a vertical line to represent the range of the day's trading (i.e. joining the high and the low) and a horizontal bar to mark the closing price. Unlike a point-and-figure chart, which has no time scale, the line-and-bar chart has one entry each day and, therefore, a horizontal time scale.

Analysing the charts

The methods for analysing both of these charts are similar, though the line-and-bar chart is more popular and therefore used to illustrate the examples. Anyone interested in investigating charting methods in depth will find a large number of detailed books on the subject, but few oil-industry men will need more than a rudimentary understanding.

Trend lines

Trend lines, as shown in Figure 9.2, are perhaps the most simple tool. They consist of a line drawn between at least three highs (for a down-

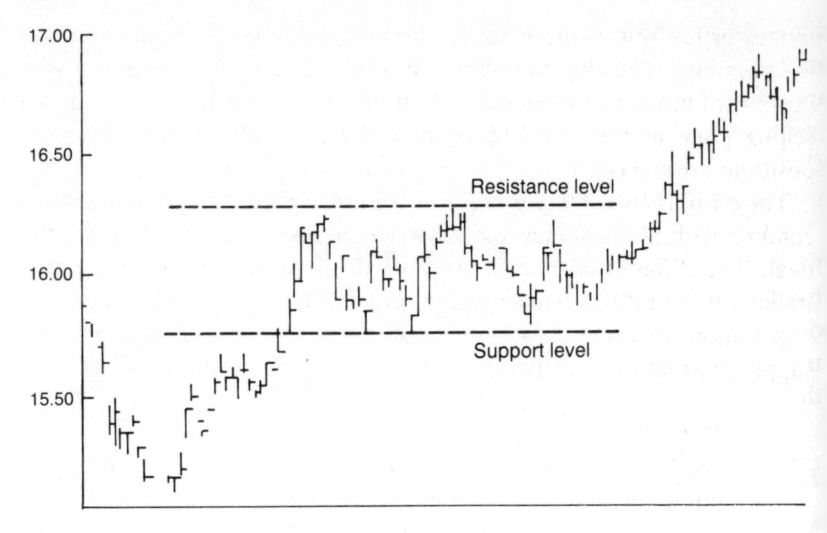

Figure 9.3 Determining support and resistance levels.

trend) or three lows (for an uptrend). While the market continues to trade on one side of the line, the trend can be expected to continue, but as soon as it closes on the other side the trend will be reversed.

While the market is in a trend, the chartist may well move in and out of the market several times, but will always play from the same side. Thus, in a downtrend, he will normally be short, though he may, from time to time, cover and then sell out again to take advantage of short-term moves.

Once the market has reversed, the chartist will close out existing positions but will not reverse them until a new trend line has been formed, giving confirmation of the change of direction. In all charting methods, the top and bottom of a move will be missed, but the trend will not.

Selecting stop levels

Another use of charts is for determining support and resistance lines, frequently used to select stop levels. These are illustrated in Figure 9.3. A support line is a horizontal (or virtually so) line drawn to join several lows while a resistance line connects highs. They therefore indicate the levels at which the market has halted previously and may well do so again. The more times a market bounces off a support or resistance line, the stronger the reaction will be when it finally breaks through. If a market closes several times at around the same level, this can also be used as a measure of support or resistance.

As soon as a support or resistance line is broken, it immediately changes its function. Thus, a broken support line becomes a resistance indication and vice versa.

Support and resistance lines are used to determine stop levels because they indicate that the market has moved out of the range previously traded. In New York particularly, this can result in a wave of selling orders triggered off as soon as the market touches a certain level. Any trader wishing to buy would therefore be advised to wait for some of these to come out and push the market down.

Double bottoms and double tops

Most of the other chart indications come from the formation of a pattern

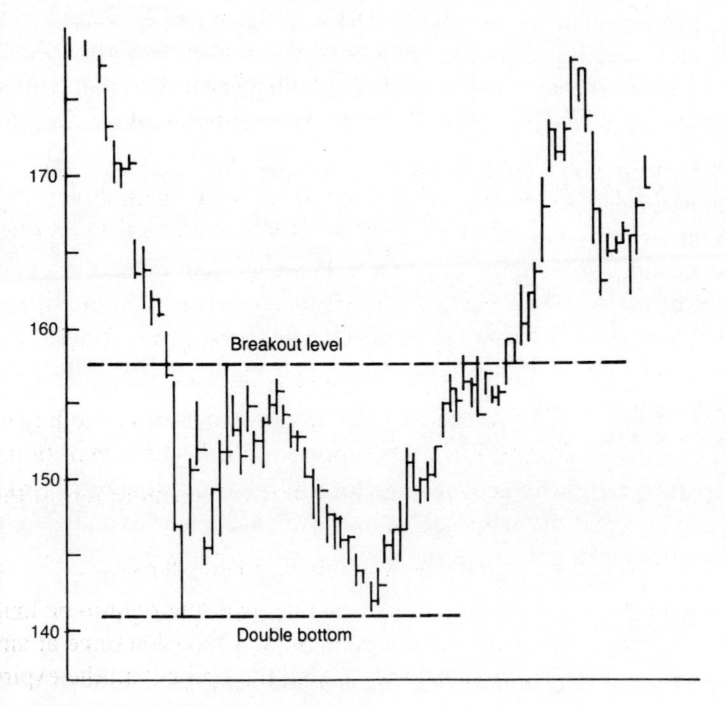

Figure 9.4 The double bottom pattern.

rather than the simple drawing of a line. The two most common of these are double bottoms and double tops, and head-and-shoulders.

Figure 9.4 illustrates a double bottom. It is made up of four separate parts:

1. A fall in price.
2. An upward reaction.
3. A second fall to around the same level as the first.
4. A second upward move.

When the second upward move takes the price beyond the first rally, a buy signal is given.

A double top is the reverse of this: with a rally, a downward reaction, a second rally and a second fall to below the level of the first. This will give a sell signal.

Head-and-shoulders

A head-and-shoulders (see Figure 9.5) requires a longer sequence of events and, despite its name, may be drawn either way up. Like a double top or bottom it results in a sell or buy signal. It is made up as follows:

1. A rally in good volume (the top of the rally becomes the left shoulder).
2. A minor recession.
3. A second high-volume rally to a level higher than the left shoulder (the head).
4. A second drop to below the level of the left shoulder.
5. Another rally, in thin volume, which fails before reaching the head (the right shoulder).
6. A further recession, breaking the neckline.

A neckline is drawn to connect the lows of the two recessions 2 and 4 (in above list) and the sell signal is given when the market falls below this line after making the right shoulder.

Moving-average charts

Perhaps the most common charting method, and the foundation of many charting systems – as well as one of the simplest to understand – is

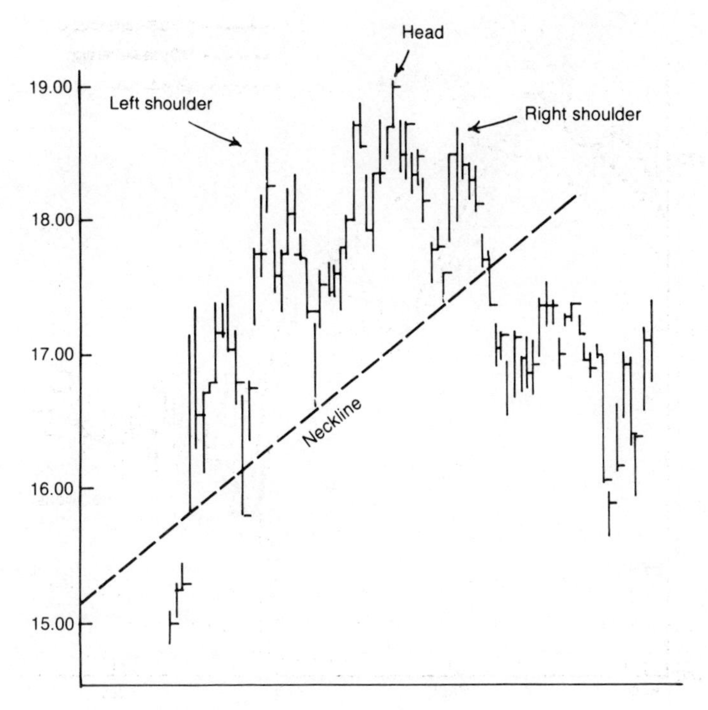

Figure 9.5 The head-and-shoulders development.

the moving-average chart. This is illustrated in Figure 9.6. A line-and-bar chart is plotted and various averages are overlaid. The most common are averages of the closing price, though highs, lows or combinations of all three may also be used. At least two, usually three, moving averages are used in conjunction with the closing price to indicate the strength or weakness of the market. Popular combinations include the 5-, 10-, 15- and 20-day averages, though 14-, 18- and 19-day are also quite common.

Whichever averages are used, the principle of reading the chart is the same. A sell signal is given when the longest-term average is above the next and so on with the close at the bottom. In the illustration the 25-day is above the 10-day, which in turn is above the 5-day which is above the close. A buy signal is given when all four are in the reverse order.

Obviously, the larger the number of days chosen for the longest-term average, the longer the chart will take to change its indication. But if the

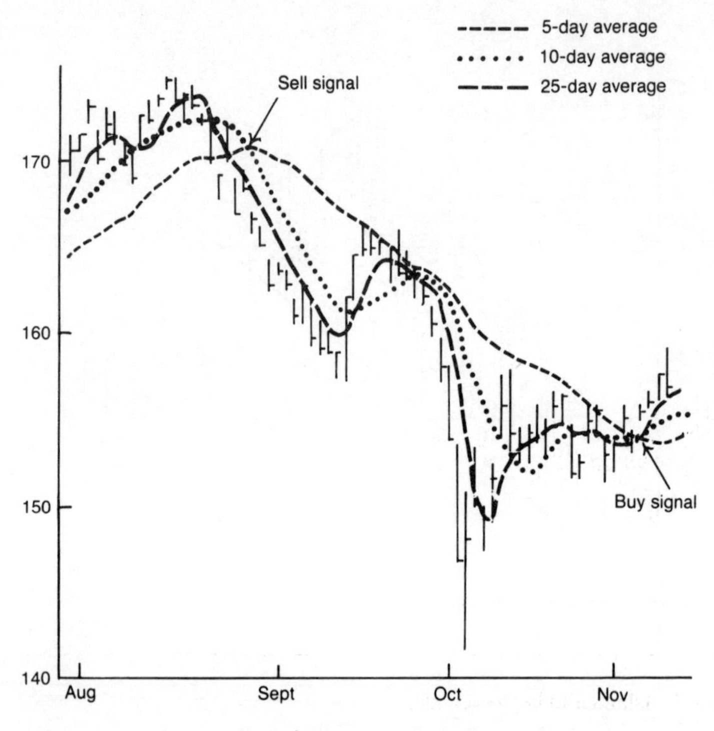

Figure 9.6 A moving-average chart.

averages are too short, they may turn before a trend has properly established itself.

The shorter-term trends are used to determine that the market is still within the trend indicated by the longer-term one. In all cases, a trader is seeking to find the optimum combination which will give indication to a trend as quickly as possible without becoming too susceptible to short-term fluctuations.

Reversal patterns

There are several reversal patterns to be found on charts, the most common being (see Figure 9.7) the island reversal and key reversal (also called an outside day reversal).

An island reversal in an upward market occurs when one day's trading has a low which is higher than the previous day's high and the

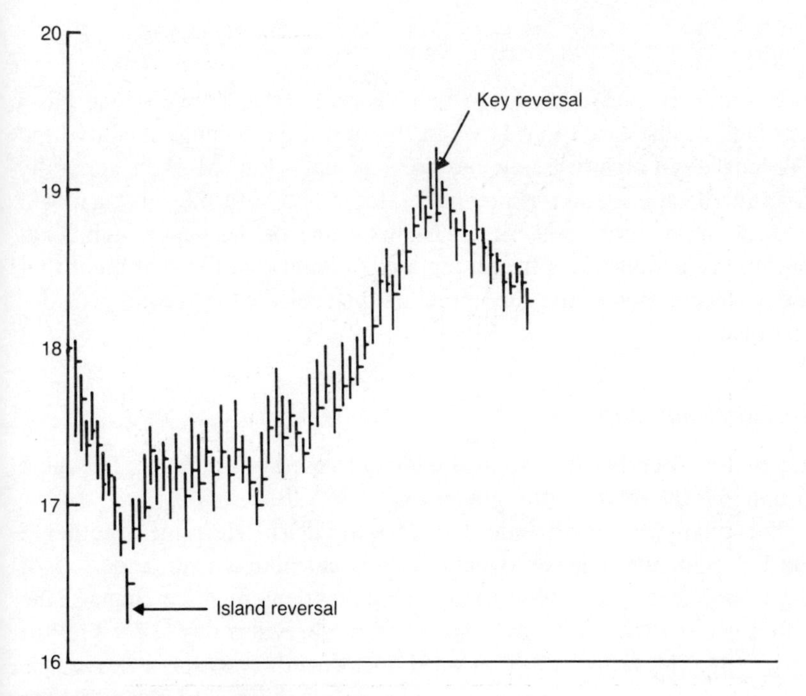

Figure 9.7 Island and key reversals.

following day's high, leaving that day's mark on the chart as an island. It tends to happen because the last of the shorts have been driven out of the market by the rising prices, leaving, in theory, no buyers to push prices further. In fact, it is not necessary for the island to be only one day's trading; it may be several days'. An island reversal in a downward market is the reverse, a day (or days) where the high is lower than the previous day's low or the following day's low.

A key reversal is somewhat similar, except instead of leaving a gap on the chart like an island reversal the apparent change of sentiment occurs during market trading hours. In a down-trending market a key reversal results from a day where a new low is seen followed by a rally to above the previous day's high and a settlement also above the previous day's high. It is indicative of the same change of sentiment as the island reversal. A key reversal of an uptrend is the reverse, a day with a new high but which settles below the previous day's low.

Gaps

A gap occurs on a chart when there is no overlap between one day's trading and the next, that is when the entire day's range is above the previous day's high or below the previous day's low. Markets generally try and fill gaps as soon as they are entered and both sides of a gap will provide support or resistance. The first edge of the gap is significant because it demonstrates the strength or otherwise of the sentiment that led to the creation of the gap in the first place while the second provides an objective.

Technical indicators

There are a number of technical indicators used by chartists. The most common is the relative strength index.

The relative strength index (RSI) is used to determine whether a market is overbought or oversold. It is calculated over a period of days, most commonly nine days. Taking settlement price change (the difference between one day's close and the previous day's) for the last nine, say, days, the total upward move and downward move are calculated. The upward move is divided by the downward move and the result taken from a hundred. Thus a total upward move of 75 cents and downward move of 10 cents would give $(100 - 75/10) = 92.5$.

An RSI over 80 is usually considered to indicate an overbought market and an RSI below 20 an oversold one. The indicator shows that there are too many buyers in an overbought market, pushing the price up too high, while in an oversold market the downward moves are being overstressed.

As with all technical indicators, a change in the fundamentals of a contract will break the rules (RSIs reached 99 on the NYMEX WTI market during early 1989) but the RSI is a useful indicator to the mood of the market. A very low RSI would indicate to a potential seller, for example, that he might want to wait a while before selling. It would not change his mind about whether to sell or not, but merely affect the timing.

Momentum indicators are also used extensively to monitor markets. There are a number of these used, probably the most common being the simple stochastic which calculates momentum from the closing price and its place in the recent trading range. It seeks to determine when a market is running out of enthusiasm for one directional move and therefore is likely to turn round.

There are also a large number of statistical and behavioural theories which have been developed over many years. These are of interest to those wanting to study charts in detail but are not for those who wish merely to have a rudimentary understanding of how the markets will be affected by charting theory.

Charts and the oil industry

It is unlikely that industry traders will become significant users of charting methods in the oil markets. Anyone with knowledge of the fundamentals is unlikely to be able to forget them and blindly follow whatever the charts are saying. None the less, all users of the market should have some understanding of what the various indications mean. Even if they do not look at charts themselves, their broker will certainly inform them of what the charts are saying and the volume of trading behind the charts makes them impossible to ignore; though ultimately, of course, the market will react to fundamentals.

There are an enormous number of 'trading systems' employed by brokers operating commodity funds. All of these are based on fairly simple charting techniques, as described here, with added refinements, mostly aimed at anticipating moves, and therefore trading, earlier than other systems. These funds have varying degrees of success and are an increasingly popular investment, particularly in the United States where tax laws are more favourable towards investors in commodities than they are in Europe.

There is no real magic about charts. They identify trends, and will therefore be unsuccessful in a flat, drifting market with no clear direction, or one with sharp fluctuations but no overall trend. By definition, they will miss the top and bottom of the market, but they should also iron out 'freak' movements in price – when, for example, the market moves sharply on a rumour which turns out to be untrue.

Glossary of terms

Actuals	The physical commodity (also called physicals).
ADP	Alternative Delivery Procedure. A procedure where delivery can be effected outside the rules of the exchange.
Arbitrage	The simultaneous purchase of futures in one market and sale of the same quantity of the same or similar product in another.
At-the-money	An option with a strike price at the current futures price.
Backwardation	The price differential between months when the nearby month is at a premium to the further one.
Basis	The price differential between the futures price and the physical price of the same product.
Bear	Someone who thinks prices will fall.
Best	An 'at best' order allows the broker to buy or sell at the best price available at his discretion.
Bid	A commitment to buy one or more lots at the specified price.
Bull	Someone who thinks prices will rise.
Call option	An option giving the buyer the right to be long of a futures contract at a specified price at any time between taking the option and the expiry date.

CFTC	The Commodities Futures Trading Commission, the regulatory body for US futures markets.
Clearing House	An agency with which all futures contracts are registered and which guarantees the financial performance of all contracts.
Contango	The price differential between months when the nearby month is at a discount to the further one.
Cross	(London and Singapore only) When a broker has buy and sell orders at the same price he will trade with himself to register the trades on the floor.
Day trade	A position opened and closed within the same trading day. A preferential commission rate may be given for such trades.
Declaration date	The date on which options expire. (Also called expiration date.)
Delta	The amount by which an option premium changes with the underlying futures price.
E.f.p.	Exchange for physicals. A physical deal priced on the futures market.
Exercise	The conversion of an option to a futures contract.
Expiration date	The date on which an option expires. (Also called declaration date.)
First notice day	The first day on which notice of intention to deliver onto the market may be made. (On all current oil contracts there is only one notice day.)
Good-till-cancelled	An order which remains in the market until it is executed or cancelled. (Normally unfilled orders are cancelled at the end of the day.)
Granter	The seller of an option.
Hedge	Taking a futures position equal and opposite to a physical position.

Implied volatility	The volatility of the market as determined from the option premium.
In-the-money	An option which would be profitable if exercised.
Intrinsic value	The amount by which an option is in-the-money.
Last trading day	The last day on which a contract month is traded on the market.
Limit	(Does not apply to the spot month.) The maximum movement allowed from the previous day's settlement.
Liquidation	The closing of an existing position on the futures market.
Long	Someone who has bought futures.
Lot	A standard contract unit.
Margin (original)	The deposit charged by the clearing house when a position is opened.
Margin (variation)	The difference in value between the price at which a contract was opened and the current market price.
Market order	An order to be filled at the prevailing market price.
Offer	A commitment to sell one or more lots at the specified price.
Open position	The number of unclosed contracts on the market.
Option	An instrument giving the buyer the right to be long or short of a futures contract at a specified price at any time between taking the option and the expiry date.
Out-of-the-money	An option which would be unprofitable if exercised.
Premium	The amount paid by the buyer of an option for the rights obtained.
Put	An option giving the buyer the right to be short

	of a futures contract at a specified price at any time between taking the option and the expiry date.
Settlement price	The price at which all futures positions are margined. It is a representative price for the close of the day's trading and is established in different ways on different markets.
Short	Someone who has sold futures.
Spot month	The nearest traded futures month.
Stop order	An order which becomes a market order as soon as the contract trades at a specified price. (Often used to limit losses, when it is called a stop loss order.)
Strike price	The price at which an option is exercised.
Spread	The differential in price between two months in the same contract or the same month in different contracts.
Straddle	Another term for a spread (also the name of an options strategy).
Switch	Another name for a spread.
Taker	The buyer of an option.
Time value	The part of an option premium reflecting the time left to expiry.
Writer	The seller of an option.

APPENDIX A

Contract specifications

International Petroleum Exchange

1. Gas oil No. 2

Quantity: 100 tonnes with volumetric delivery of 118.35 cu. m.

		Test method
Density at 15°C kg/litre	0.855 max 0.820 min	ASTM D 1298 (vacuum)
Distillation, % volume		
evaporated at 250°C	64 max	
350°C	85 min	
Colour	1.5 max	ASTM D 1500
Flash point		
Pensky Martens 'Cloud Cup', °C	60 min	ASTM D 93
Total sulphur, % weight	0.3 max	ASTM D 1552
Kinematic viscosity		
centistokes at 20°C	6.0 max	ASTM D 445
Cloud point, °C		
April–September	+2.0 max	ASTM D 2500
October–March	−2.0 max	
Cold filter plugging point, °C		
April–September	−7.0 max	IP 309
October–March	−11.0 max	
Oxidation stability, % weight	3.0 max	ASTM D 2274
Cetane index	45 min	ASTM D 976/80
Sediment, % weight	0.05 max	ASTM D 473
Water, % volume	0.05 max	ASTM D 95

Delivery: f.o.b. Amsterdam/Rotterdam/Antwerp.
Quotation: the price is quoted in US dollars and cents per tonne.
Minimum fluctuation: 25 cents/tonne.

Maximum fluctuation: the maximum move permissible in one day is $15 per tonne above or below the previous day's settlement price. If the market moves to limit trading will cease for half an hour, reopen and continue without limit for the rest of the day. (No limit on the spot month.)

Trading hours: 09.15–17.24.

Last trading day: the spot month shall cease trading at noon on the third business day before the thirteenth of the month.

2. Brent Blend

Quantity: 1,000 barrels.

Specification: current pipeline export quality Brent Blend for delivery at the Sullom Voe delivery area.

Delivery: cash settlement at the average IPE Brent daily index price for the last trading day.

Quotation: the price is quoted in US dollars and cents per barrel.

Minimum fluctuation: 1 cent/barrel.

Maximum fluctuation: no limit.

Trading hours: 09.30–17.30.

Last trading day: the tenth of the month or last business day before.

3. Heavy fuel oil

Quantity: 100 tonnes.

Quality is defined as: high sulphur fuel oil as traded f.o.b. barges in the Amsterdam, Rotterdam, Antwerp area with a typical sulphur level of 3.5 per cent weight and cargoes c.i.f. North-West Europe with a maximum sulphur level of 3.5 per cent weight.

Delivery: cash settlement against the IPE fuel-oil index.

Quotation: the price is quoted in US dollars and cents per tonne.

Minimum fluctuation: 10 cents/tonne.

Maximum price fluctuation: there is no limit on the daily price move.

Trading hours: 09.15–17.30.

Last trading day: trading shall cease at close of business three business days before the thirteenth day of the delivery month.

New York Mercantile Exchange

4. No. 2 heating oil

Quantity: 42,000 US gallons

		Test method
Gravity, API	30 min	ASTM D 287
Flash point, °F	130 min	ASTM D 93

Kinematic viscosity, centistokes at 100°F	2.0 min 3.6 max	ASTM D 445
Water and sediment, %	0.5 max	ASTM D 1796 or D 2709
Pour point, °F	0 max	ASTM D 97
Distillation, °F		
10% evaporation	480 max	ASTM D 86
90% evaporation	640 max	
End point	670 max	
Sulphur, % weight	0.2 max	ASTM D 129 or D 1552
Colour	2.5 max	ASTM D 1500

Delivery: ex seller's facility in New York Harbor.
Quotation: the price is quoted in US cents per gallon.
Minimum fluctuation: 0.01 cents/gallon.
Maximum fluctuation: the maximum move permissible in one day is 2 cents/gallon above or below the previous day's settlement price. Each day the market settles at limit in any month, the limit shall be increased by 1 cent/gallon to a maximum of 4 cents/gallon. Each day the market does not settle at limit in any month, the limit shall be reduced by 1 cent/gallon to a minimum of 2 cents/gallon. (No limit on the spot month.)
Last trading day: the spot month shall cease trading at the close of business on the last business day of the month preceding delivery.
Trading hours: 09.50–15.10.

5. Unleaded regular gasoline

Quantity: 42,000 US gallons

		Test method
Gravity, API	52 min	ASTM D 287
Colour	Orange or bronze to meet US Surgeon General's requirements	
Corrosion, 3 hours at 122°F	1 max	ASTM D 130
Lead, g/gallon	0.2 max	ASTM D 3341 or equivalent
Doctor test	negative	ASTM D 484
or		
Mercaptan sulphur	0.002 max	ASTM D 3227
Octane RON	Report min 91.5	ASTM D 2699
MON	Report	
(RON + MON)/2	89.0 min	

Reid vapour pressure, p.s.i.	Months		
	December, January, February	14.5 max	ASTM D 323
	March, April, October, November	13.5 max	
	May, June, July, August, September	11.5 max	

Distillation Northern Grade	Class	
December, January, February	E	ASTM D 86
March, April, October, November	D	
May, June, July, August, September	C	

	B	C	D	E
10% evaporation °F max	149	140	131	122
50% evaporation °F min	170	170	170	170
50% evaporation °F max	345	340	235	230
90% evaporation °F max	374	365	365	365
End point °F max	430	430	430	430

Delivery: ex seller's facility in New York Harbor.

Quotation: the price is quoted in US cents per gallon.

Minimum fluctuation: 0.01 cents/gallon.

Maximum fluctuation: the maximum move permissible in one day is 2 cents/gallon above or below the previous day's settlement price. Each day the market settles at limit in any month, the limit shall be increased by 1 cent/gallon to a maximum of 4 cents/gallon. Each day the market does not settle at limit in any month, the limit shall be reduced by 1 cent/gallon to a minimum of 2 cents/gallon. (No limit on the spot month.)

Trading hours: 09.30–15.10.

Last trading day: the spot month shall cease trading at close of business on the last business day of the month preceding delivery.

6. Light sweet crude oil

Quantity: 1,000 barrels

		Test method
Gravity, API	34 min 45 max	ASTM D 129
Sulphur, % weight	0.5 max	ASTM D 287

Viscosity, seconds Saybolt universal	550 max	
Reid vapour pressure, p.s.i. at 100°F	9.5 max	ASTM D 323
Basic sediment, water, etc. volume %	1.0 max	ASTM D 473
Pour point, °F	50 max	

Stream designations: Domestic West Texas Intermediate
Mid Continent Sweet
Low Sweet Mix (Scurry Snyder)
New Mexican Sweet
North Texas Sweet
Oklahoma Sweet
South Texas Sweet

Foreign UK: Brent Blend
Nigeria: Brass Blend
Bonny Light
Norway: Ekofisk
Tunisia: Zarzaitine/El Borma
Algeria: Saharan Blend

Delivery: f.o.b. Cushing, Oklahoma.

Differentials: differentials as fixed by the exchange shall be applied to crude oils of higher gravity and higher or lower sulphur content than the par of 35 API and 0.5 per cent respectively.

Quotation: the price is quoted in US dollars and cents per barrel.

Minimum fluctuation: 1 cent/barrel.

Maximum fluctuation: the maximum move permissible in one day is $1 per barrel above or below the previous day's settlement price. Each day the market settles at limit in any month, the limit shall be increased by 50 cents/barrel to a maximum of $2 per barrel. Each day the market does not settle at limit in any month, the limit shall be reduced by 50 cents/barrel to a minimum of $1 per barrel. (No limit on the spot month.)

Trading hours: 09.45–15.10.

Last trading day: the spot month shall cease trading at the close of business on the third business day preceding the twenty-fifth day of the month preceding delivery.

7. Heavy fuel oil

Quantity: 1,000 US barrels.

Quality: heavy fuel oil with a maximum sulphur content of 1 per cent weight.

Delivery: Houston Ship Channel, West of Lynchburg, Texas.

Quotation: the price is quoted in US dollars and cents per tonne.

Minimum fluctuation: 1 cent/barrel.

Maximum price fluctuation: the maximum move permissible in one day is $1.00 per barrel above or below the previous day's settlement price. Each day the market settles at limit in any month the limit will be increased by 50 cents per barrel to a maximum of $2.00 per barrel. Each day the market does not settle at limit in any month the limit shall be reduced by 50 cents per barrel to a minimum of $1.00 per barrel.

Trading hours: 09.45–15.10.

Last trading day: trading shall cease at close of business on the last business day of the month preceding the delivery month.

Singapore International Monetary Exchange

8. Fuel oil

Quantity: 100 tonnes

		Test method
Density at 15°C	0.991 max	ASTM D 1298/IP 60
Sulphur, %	4.0 max	ASTM D 2622
Kinematic viscosity, centistokes at 50°C	180 max	ASTM D 445/IP 71
Flash point, °C	60 min	ASTM D 93/IP 34
Pour point, °C	24 max	ASTM D 97/IP 15
Ash, %	0.15 max	ASTM D 482/IP 4
Water by distillation, %	0.5 max	ASTM D 95
Conradsen carbon, %	18 max	ASTM D 189/IP 13
Vanadium, p.p.m.	400 max	ASTM D 1548/IP 288
Aluminium, p.p.m.	30 max	IP 377
Sodium, p.p.m.	160 max	ASTM D 1318/IP 288
Hot filtration, %	0.10 max	IP 375

Delivery: f.o.b. Singapore Port.

Quotation: the price is quoted in US dollars and cents per tonne.

Minimum fluctuation: 10 cents/tonne.

Maximum price fluctuation: if the price moves $5 per tonne above or below the previous day's settlement price the limit will be extended to $10 per tonne after half an hour. If the market settles at the $10 per tonne limit for two successive days there shall be no limit on the third day. (No limit on the spot month.)

Trading hours: 09.00–18.00.

Last trading day: the contract will cease trading on the last business day of the preceding month.

APPENDIX B

Costs of futures trading

There are three basic components in futures trading: initial margins (or deposits), variation margins and commission to the broker. Of these, the first is fixed by the relevant exchange, the second dependent on the price and the third entirely negotiable with the broker.

Initial margins are payable as soon as a position is opened. They are varied from time to time by the clearing systems, according to the degree of volatility in the price. In early 1989 they stood at $1,000 per lot on the IPE, $2,000 per lot on NYMEX and $500 per lot on SIMEX. These margins are equivalent to around 5–10 per cent of the contract value and are held by the clearing house to enable the clearing system to guarantee performance of the contract. In effect, they are supposed to cover against a daily price move because variation margins are always effectively one day behind.

On most exchanges, initial margins are increased sharply towards the expiry of a contract: for example, on the IPE they are doubled five days before expiry and doubled again three days later and on NYMEX they are increased by $1,000 a month before expiry and a further $2,000 five days before expiry.

Spread trading within one exchange, either between two months on one market or between two markets on one exchange, usually attracts lower initial margins. Spreads involving the spot month are sometimes excluded from this concession or attract additional payments as if they were outright trades.

Variation margins are payable daily and represent the full difference between the market value of the contract at the settlement price and the price at which it was taken out. For example, if a trader is long one lot (1,000 barrels) of Brent at $18.00 and the price falls to $17.50, he is required to pay the clearing house (via his broker) the full 50 cents/barrel. If the price rises to $18.50 the clearing house will pay him the full 50 cents/barrel.

Index